The Textile Institute

Textile Progress
Volume 31
Number 1/2

The Science of Clothing Comfort

A critical appreciation of recent developments by Y. Li

Edited by J.M. Layton

THE TEXTILE INSTITUTE
INTERNATIONAL
Fourth Floor, St. James's Buildings, Oxford Street, Manchester M1 6FQ, UK

ISSN 0040 5167
ISBN 1 870372 24 7

Printed by Alden, Oxford, UK

CONTENTS

THE SCIENCE OF CLOTHING COMFORT

Y. Li

1. CLOTHING COMFORT

1.1 Consumer Trends in the 90s

Modern consumers are interested in clothing that not only looks good, but also feels good. They would like their clothing to coincide with their chosen attitudes, roles, and images. It has been identified, by both natural and synthetic fiber marketers, that consumers are increasingly involving more than their visual sense and are allowing touch, smell, intuition, and emotion to influence their decisions. As a result, greater importance is being attributed to the shopping and wearing experience. Interest is growing in better feeling fabrics. Comfort is being reinforced as a key parameter in clothing [1].

Comfort has been identified by major fiber marketers as one of the key attributes for consumers' desirability on apparel products in all markets. However, retailers and processors feel some uncertainty with regard to consumer requirements. Under the heading *comfort*, the problem of prickle, especially in knitwear, is universally recognized, and lightweight, fluid fabrics are seen as highly desirable. Consumers find *luxurious softness* somewhat less appealing than do processors and retailers. Also, it has been shown that consumers' requirements on comfort are changing with products and wear situations. The underlying patterns and reasons behind these are still myths that need scientific investigation.

In the 90s, the apparel market is highly competitive. To meet, and even exceed, consumers' needs and expectations becomes essential if enterprises in the textile and clothing industry are to succeed in the market place. Synthetic fiber manufacturers have made a successful comeback through the sportswear route, where they have emphasis on comfort, movement, and performance. By addressing consumers' needs, and using the growing interface between sportswear and fashion, synthetic fiber manufacturers have taken half of the fiber consumption market. Clearly, understanding and satisfying consumers' needs and wants towards apparel products is crucial for the long-term survival of any enterprise in the rivalry of the market place.

Comfort is a fundamental and universal need for consumers. As consumers ourselves, everything we do can be considered to be an effort to improve our level of comfort in life. Clothing and textile products are essential materials that we use everyday to obtain physiological and psychological comfort and, more fundamentally, to ensure physical conditions around our body suitable for survival. Therefore, research on clothing comfort has fundamental meanings for the survival of human beings and improvement of the quality of our life. From the viewpoint of the business management of textile enterprises, clothing comfort research has substantial financial implications in the effort to satisfy the needs and wants of consumers in order to obtain sustainable competitive advantages in modern consumer markets.

1.2 Definition of Comfort

Comfort is a complex and nebulous subject that is very difficult to define. Fourt and Hollies surveyed the literature and found that comfort involves thermal and non-thermal components and is related to wear situations such as working, non-critical and critical conditions [2]. The physiological responses of the human body to a given combination of clothing and environmental conditions are predictable when the system reaches a steady state. It can be calculated from a knowledge of easily measured factors, such as the thermal resistance and moisture resistance of the clothing, the climatic conditions, and the level of physical activity. This is the traditional investigational area in clothing comfort research, on which a large amount of work has been published and applied to solve practical problems. For instance, the thermal insulation value, *clo*, has been widely used for designing and classifying military uniforms [3], and calculating thermal comfort indices for indoor air conditioning [4–6].

As clothing is directly in contact with the human body, it interacts with the body continuously and dynamically during wear, which stimulates mechanical, thermal, and visual sensations. This has been termed *sensory comfort*, which is a relatively new area in clothing comfort research.

Slater [7] defined comfort as 'a pleasant state of physiological, psychological, and physical harmony between a human being and the environment'. Slater identified the importance of environment to comfort and defined the three types named. Physiological comfort is related to the human body's ability to maintain life, psychological comfort to the mind's ability to keep itself functioning satisfactorily with external help, and physical comfort to the effect of the external environment on the body.

It has been recognized for a long time that it is difficult to describe comfort positively, but discomfort can be easily described in such terms as *prickle, itch, hot,* and *cold.* Therefore, a widely accepted definition for comfort is 'freedom from pain and from discomfort as a neutral state' [8]. Further, the psychological and physiological states have a number of aspects:

- Thermophysiological comfort – 'attainment of a comfortable thermal and wetness state; it involves transport of heat and moisture through a fabric'.
- Sensorial comfort – 'The elicitation of various neural sensations when a textile comes into contact with skin'.
- Body movement comfort – 'Ability of a textile to allow freedom of movement, reduced burden, and body shaping, as required'.
- Aesthetic appeal – Subjective perception of clothing to the eye, hand, ear, and nose, which contributes to the overall well-being of the wearer [8].

In all these definitions, there are a number of essential components:

- Comfort is related to subjective perception of various sensations.
- Comfort involves many aspects of human senses, such as visual (aesthetic comfort), thermal (cold and warm), pain (prickle and itch), and touch (smooth, rough, soft, and stiff).
- The subjective perceptions involve psychological processes in which all relevant sensory perceptions are formulated, weighed, combined, and evaluated against past experiences and present desires to form an overall assessment of comfort status.
- Body–clothing interactions (both thermal and mechanical) play important roles in determining the comfort state of the wearer.
- External environments (physical, social, and cultural) have great impact on the comfort status of the wearer.

This suggests that comfort is multidimensional and complex. Subjective perception of comfort involves complicated processes in which a large number of stimuli from clothing and external environments communicate to the brain through multi-channels of sensory responses to form subjective perceptions.

1.3 The Human-clothing System

Clothing is an integral part of human life, and has a number of functions: adornment, status, modesty, and protection. Clothing, by means of the latest fashion and aesthetic appeal, can provide the wearer with adornment that gives him or her the mental comfort of looking good. Well-fitting and luxurious clothing can enhance the status of the wearer, which gives him or her a feeling of satisfaction. Clothing can also provide the function of modesty by giving mental comfort in covering the body sufficiently to meet the standards of society and cover physical flaws. However, the primary role of clothing is as a layer or layers of barriers that protect the body against unsuitable physical environments. This protection fulfils a number of functions: maintaining the right thermal environment to the body that is essential for its survival; and preventing the body from being injured by abrasion, radiation, wind, electricity, chemical, and microbiological toxic substances.

These traditionally classified functions of clothing indicate clearly that it plays very important roles at the interface between a human body and its surrounding environment in determining the subjective perception of comfort status of a wearer. To understand how the subjective perception of comfort is derived, we can consider human-clothing as an open system that is always in a state of dynamic interaction with its surrounding environment in physical, sensory, psychological, and informational means (see Fig. 1). In this system, there are a number of processes occurring interactively which determine the status of comfort of a wearer:

- Physical processes in the clothing and the surrounding environment, such as the heat and moisture transport in the clothing, mechanical interactions between the clothing and the body, and reflection and absorption of light by the clothing, which provide physical stimuli (or signals) to the body.
- Physiological processes in the body, such as the thermal balance of the body, and its thermoregulatory responses and dynamic interactions with the clothing and the environment, which determine the physiological status of the body and its survival under critical conditions.
- Neurophysiological processes, i.e. the neurophysiological mechanisms of the sensory reception system of the body in the skin, eyes, and other organs, by which sensory signals are formulated from the interactions of the body with the clothing and surrounding environments.

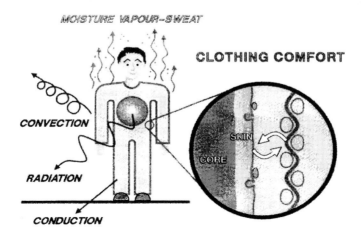

Fig. 1   The human–clothing–environment system

- Psychological processes, i.e. the processes of the brain which form subjective perception of sensory sensations from the neurophysiological sensory signals and then formulate subjective overall perception and preferences by evaluating and weighing various sensory perceptions against past experiences and internal desires.

These four types of processes are occurring concurrently. The physical processes in the environment and clothing follow the laws of physics, which determine the physical conditions for the survival and comfort of the body. The thermoregulatory responses of the body and the sensory responses of skin nerve endings follow the laws of physiology. The thermoregulatory and sensory systems respond to the physical stimuli from clothing and environment to ensure appropriate physiological conditions being met for the survival of the body, and to inform the brain of various physical conditions that influence the status of comfort. The psychological processes, of which we have least understanding, are the most complex. The brain needs to formulate subjective perceptions from the sensory signals from the nerve endings, and to evaluate and weigh these sensory perceptions against past experiences, internal desires, and external influences. Through these processes, the brain formulates subjective perception of overall comfort status, judgments, and preferences.

On the other hand, the psychological power of the brain can influence the physiological status of the body through various means such as sweating, blood-flow justification, and shivering. These physiological changes will change the physical processes in the clothing and external environment. These four types of processes interact with each other dynamically to determine the comfort status of the wearer at any specific moment. Therefore, comfort status is the subjective perception and judgment of a wearer on the basis of integration of all of these physical, physiological, neurophysiological, and psychological processes and factors.

1.4 Scope of Clothing Comfort Research

It is obvious that development of an understanding of all the individual processes outlined, and their interactions, is essential for obtaining a knowledge of comfort. Investigation of the mechanisms involved in these processes is the fundamental research that establishes the theoretical framework of the science of clothing comfort.

Following the theme defined, the scope of clothing comfort research needs to include four essential areas: physics, physiology, neurophysiology, and psychology of comfort. Significant progress has been made in the past few decades in all four areas by researchers around the world from many disciplines; these are discussed and reviewed critically in Sections 2 to 7.

Further, researchers have made great efforts to integrate all the knowledge on the physical, physiological, neurophysiological, and psychological processes to predict comfort performance of clothing during wear. In Section 8, the predictability of clothing comfort performance is discussed.

Finally, the industrial application of comfort research is another area of extreme importance. To improve the quality of life and the survival of human beings long-term is the final goal of clothing comfort research. Only through effective practical applications by industrial enterprises in their efforts to improve profits and survival chances of their businesses, can clothing comfort research be kept alive and its final goals be achieved. The final section of this *Textile Progress* focuses on the issues of how to utilize comfort research as an effective tool to obtain sustainable competitive advantages for enterprises in the textile and clothing industry.

2. PSYCHOLOGY AND COMFORT

2.1 Perception of Comfort

Comfort, in the end, is the psychological feeling or judgment of a wearer who wears the clothing under certain environmental conditions. Pontrelli developed a *Comfort's Gestalt* [9] in which the variables influencing the comfort status of a wearer were listed comprehensively. The variables were classified into three groups: physical variables of the environment and the clothing; psycho-physiological parameters of the wearer; and psychological filters of the brain. The Gestalt indicates that the comfort status of a wearer depends on all these variables and their interactions.

In consideration of the formulation processes of the subjective perception as discussed in Section 1.3, the flow chart for the subjective perception of comfort can be drawn as follows:

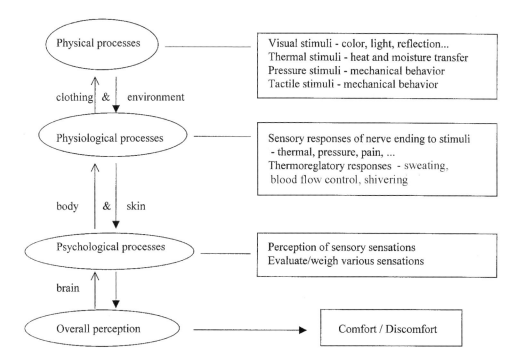

This chart illustrates the processes of how the subjective perception of overall comfort is formulated. The physical processes provide the signals or stimuli to the sensory organs of the human body, which will receive them, produce neurophysiological impulses, send these to the brain, and take action to adjust sweating rate, blood flow, and sometimes heat production by shivering. The brain will process the sensory signals to formulate subjective perception of various individual sensations, and further evaluate and weigh them against past experience and desires, which is influenced by many factors such as physical, environmental, social and cultural surroundings, and state of being.

The psychology of comfort is the study of how the brain receives individual sensory sensations, and evaluates and weighs the sensations to formulate subjective perception of overall comfort and preferences which become our wear experience and influence our further purchase decisions.

2.2 Psychological Research Techniques

Human perception of clothing and external environment involves all the relevant senses and has formed a series of concepts that we use to express these perceptions to each other. To understand the psychological processes we need to measure these perceptions in subjective ways. A subjective measure is the direct measure of the opinion of a person, which is the only factor of interest in carrying out the measurements. Since there are no physical instruments to measure what a wearer is thinking or feeling objectively, the only way to obtain the subjective perceptions is by use of psychological scaling. With psychological scaling, the process of making judgments is based on the scales of individual words or language that we collect from experience and share with peers throughout life.

Slater (1986) pointed out a number of problems with subjective measurements. Firstly, measurements rely completely on the honesty of human subjects. Secondly, there exist wide variations in subjective opinions in human beings, which demand a large number of measurements to obtain satisfactory precision. Thirdly, there are great difficulties in carrying out statistical analysis on subjective data because subjective answers are not real numbers, and the mental calibration used by each respondent may not be the same. Finally, there are inconsistencies in subjective data, as the opinion of individual respondents is influenced by a large number of psychological, physiological, social, and environmental factors [10].

Despite the difficulties involved, scientific psychology has been developed for over 100 years to study the behavior of humans [11]. A great deal of work has been done in the field of psychological scaling, which has developed psychological laws, experimental techniques, and mathematical methods to handle the data from subjective responses [12]. Many researchers have applied the psychological scaling techniques to study clothing comfort.

Hollies (1977) summarized six essential elements in psychological scaling:

- Commonly recognized attributes to measure.
- Language (terms) to describe these attributes.
- Assignment of a scale to indicate the level of attributes.
- A rating panel to apply the rating scale to attribute measurement.
- Appropriate data handling.
- Comparison of results from psychological scaling and objective measurement of the same attributes [13].

This indicates that the psychology of clothing comfort involves a number of research techniques; these will be discussed in detail in subsequent sections.

2.3 Comfort Sensory Descriptors

Sensations generated from clothing depend largely on the various combinations of human activities and environmental conditions experienced during day-to-day living. Researchers have identified many commonly recognized attributes of clothing related to comfort, involving thermal, moisture, tactile, hand, and aesthetic experiences. This type of identification has greater input from the view of experts. On the other hand, it is important to know whether there are some commonly recognized comfort attributes of clothing among ordinary consumers, and what they are if they exist. This can be viewed from the processes of how sensory descriptors were obtained.

Hollies found that strong sensations were experienced when mild or heavy sweating occurred, and during modest excursions of warming or chilling following the inception of sweating [14]. By repeating experiments, Hollies *et al.* obtained a list of sensory descriptors that were generated by asking the participants to describe the sensations they experienced. The list of sensations included the descriptors *snug, loose, heavy, lightweight, stiff, staticky, non-absorbent, cold, clammy, damp, clingy, picky, rough, and scratchy* [15]. Each participant had the option to use any of these descriptors and could put in additional descriptors as experienced. These sensory descriptors were repeatedly produced by participants of wear trials conducted over many years [13–17].

In a study of fabric hand, Howorth and Oliver asked 25 participants to rank 27 fabrics and describe their reasons. Twenty-one descriptive terms and their frequency of use were obtained. Through factor analysis, they derived seven descriptors for fabric hand: *smoothness, softness, coarseness, thickness, weight, warmth, and stiffness* [18,19].

For evaluating men's winter suiting fabrics, David *et al.* generated lists of 'bipolar descriptors' by discussing them with each judge. The descriptors from all the judges were collated and listed, and were then associated with the 'Standard Definitions of Terms Relating to Textiles'. However, each individual judge had the choice of using his/her own list of descriptors. For each judge, an individual list of fourteen bipolar descriptors was produced. After eliminating the pairs of words that did not give useful contribution by analyzing the data from subjective evaluation, seven pairs of descriptors were identified: *coarse–fine, stiff–pliable, rough–smooth, harsh–soft, cool–warm, hard–soft, and rustly–quiet* [20].

In developing the methodology for evaluation of fabric handle, Kawabata and Niwa generated sensory descriptors by letting a panel of expert judges (the Hand Evaluation and Standardization Committee) judge fabric handle and asking them the reasons for their decisions. They identified terms such as KOSHI (*stiffness*), NUMERI (*smoothness*), and SHARI (*crispness*) as 'primary-hand' expressions [21].

In 1998, Li [22] carried out an investigation on the psychological sensory responses to clothing of consumers living in different countries. A survey was conducted in three countries: Britain, China, and USA. Twenty-six sensory descriptors were selected: *snug, loose, stiff, lightweight, staticky, nonabsorbent, sticky, heavy, cold, damp, clammy, clingy, picky, rough, scratchy, cool, hot, soft, warm, wet, prickly, itchy, chill, sultry, tickling, and raggy.* Altogether, 465 observations were made. Using analysis of variance and non-parametric analysis of differences, it was found that the ratings of most of the sensory descriptors were significantly different between three types of clothing: summer wear, winter wear, and sportswear, at $p < 0.01$ level. Differences in the ratings of most sensory descriptors were significantly different between Chinese and British respondents for summer wear, but not for winter wear and sportswear. No significant differences in ratings of the sensory descriptors were found between male and female respondents.

In his work on Personal Construction Theory, Kelly suggested that human participants have the ability to be specific, draw on an internal concept of a particular type of garment from their memory, and generate specific criteria to describe the garment [23]. On the basis of this theory, Fritz argued that consumers have their internal scales and concepts in evaluating fabric quality. Consumers themselves know best and they are capable of making objective, quantitative, and repeatable assessments of their sensations. Researchers should try to discover the consumers' desires in the performance of products. Therefore, sensory descriptors should be derived from consumers instead of experts or researchers. Fritz reported the usage of a repertory semantic differential grid to define product attributes using descriptive adjectives, by focus group study [24]. For example, the polar pairs of descriptors for toweling fabrics include *soft–harsh, smooth–rough, cool–hot, light–heavy, fine–coarse, crisp–limp, clammy–absorbent, natural–synthetic, sheer–bulky, clingy–flowing, crushable–resilient, lacy–plain, drapable–rigid, scratchy–silky,* and *stiff–soft.*

Fritz described the procedures as: (i) organize groups of 10 to 40 participants; (ii) present the group with the concept of a product to be investigated; (iii) encourage each participant to write down as many descriptors relating to the product as possible; (iv) hold group discussions to generate further descriptors; (v) achieve a commonly agreed list of descriptors in the group for inclusion in the semantic grid; and (vi) refine, clarify, and agree upon the precise meaning of each polar pair within the group [24].

Of the sensory descriptors from these independent studies, there are commonly recognized attributes and languages to describe the attributes related to clothing comfort. These sensations may be expressed in different languages and there are difficulties in interpreting the sensory descriptors from one language to another with exactly the same meaning. However, it is obvious that there are a number of dimensions in these sensory descriptors to describe our sensory experiences that are related to thermal, mechanical, and fabric surface stimuli, implying that the study of human comfort sensations has universal implications.

2.4 Psychophysics

In 1860, Fechner originated psychophysics to describe the mathematical relationship between the conscious experience of a sensation and an external physical stimulus [25]. His philosophy was that if we know the mathematical form of the psychophysical relation between a physical variable and its corresponding sensation, we can measure mental attributes by measuring their physical correlates. Therefore, psychophysics is about the measurement of the strength of internal sensations, which can be broadly defined as the quantification of sensory experience. This has two aspects of indication: (i) the assessment of human powers of signal detection and sensory discrimination, and (ii) the calibration of subjectively perceived intensities and other parameters of stimulation.

In 1834, Ernst Weber proposed that the thresholds (i.e. the just noticeable differences) of stimulus (ΔS_p) are proportional to the magnitude of stimulus, S_p. This is known as Weber's law, and can be expressed as:

$$\Delta S_p / S_p = K$$

where K is a constant indicating the power of a human being to detect signals and to discriminate sensations. This law holds for many stimulus attributes down to about the absolute threshold, which is the smallest magnitude of stimulus that can be perceived [25].

Fechner (1860) proposed using 'just noticeable difference' as a unit to measure internal sensation. Fechner assumed that sensation R_s increases as the logarithm of the physical stimulus magnitude, S_p; this is called Fechner's law and can be described as:

$$R_s = k \log S_p$$

where k is a constant determined by the stimulus threshold, which represents the lowest physical value eliciting a sensation, and the differential threshold providing a subjective unit of sensory intensity. This law proposes that sensation increases in arithmetic steps as the physical stimulus is increased in logarithmic steps [25] [26]. Fechner's law is internally related to Weber's Law. If Weber's law applies to the stimulus attribute in question, and the thresholds in sensation are equal, then sensation increases as the logarithm of the physical stimulus magnitude.

Stevens (1953) developed a method of magnitude estimation, as an experimental procedure to investigate the relationship between subjectively perceived intensity and physical stimulus strength. This method was applied to a very large number of different stimulus attributes. The results from each attribute conform roughly to an equation of the form:

$$R_s = a\, S_p^b$$

where *a* is a scale factor and *b* an exponent characteristic of the attribute. This equation is known as Stevens' power law [25].

These psychophysical laws indicate that there is an essential distinction between the physical stimulus and the sensation that one experiences. Weber's law and Fechner's law play some fundamental role in sensory discrimination in terms of the ability to distinguish one stimulus from another, but fail to provide a basis for measuring sensation. Stevens' law proposes a power relation between physical stimulus magnitude and internal sensation which provides a 'direct' measurement of sensation in sensory judgment processes [25].

2.5 Scales of Measurement

Psychological scaling is a type of measurement that consists of assigning *numbers* to characteristics of objects or events, according to rules which reflect some aspects of reality. In social sciences and marketing research, psychological scaling has been widely used to obtain consumers' opinions and study their attitudes and preferences. The term *number* here does not always correspond to the 'real' numbers that are obtained from objective measurement in physical means. The *numbers* cannot necessarily be added, subtracted, divided, or multiplied. They are used as symbols to represent certain characteristics of the object. The nature of the meaning of the numbers depends on the nature of the characteristics and the rules specifying how the numbers are assigned to the characteristics to be measured. These rules are arbitrary, not a result of undeniable natural law [27].

The rules governing how to assign numbers constitute the essential criteria for defining each scale. There are four types of numbers or scales of measurements: *nominal, ordinal, interval, and ratio*. Moving from *nominal* to *ratio* scales, the rules becomes more restrictive and the kinds of arithmetic operations for which the numbers can be used are increased.

Nominal scales consist of numbers used to categorize objects. A nominal number serves as a label for a class category. For instance, we can assign 0 to male and 1 to female. The number 1 does not imply a superior position to the number 0. The rule for nominal scales is that all members of a class have the same number and no two classes have the same number. The only arithmetic operation that can be performed on nominal data is the count in each category. Nominal numbers cannot meaningfully be added, subtracted, multiplied, and divided.

Ordinal scales comprise numbers or other symbols used to rank objects according to their characteristics and their relative position in the characteristics. Ordinal data indicate the relative position of objects on certain characteristic scales but not the magnitude of the differences between the objects. A mode or median may be used, but not a mean. Non-parametric statistics can be applied to ordinal data.

Interval scales consist of numbers used to rank objects in such a way that numerically equal distances on the scale represent equal distances in the characteristics being measured. But both zero and the unit of measurement are not fixed and are arbitrary. Therefore, interval data can indicate both the relative position of objects and the magnitudes of differences between the objects on the characteristics being measured. The entire range of statistics can be applied to interval scales.

Ratio scales represent numbers used to rank objects such that numerically equal distances on the scale represent equal distances of the characteristics measured and have a meaningful zero. Like interval scales, entire ranges of statistics can be applied to ratio data.

The description and applicable methods of analysis of the four types of psychological scales are summarized in Table 2-1. In clothing comfort research, all four types of psychological scales have been applied. Nominal scales have been used to code subjects such as gender, age, and place of living. Ordinal scales have been used to obtain the rankings of fabrics or garments in consideration. The most frequently used scales are the interval scales, which have been widely used to obtain the perception of various attributes of clothing and which will be discussed in detail in the section following. Ratio scales are mainly applicable to the data generated from physical instruments.

Table 2-1
Types of Psychological Scales

Scale	Rules	Usage	Applicable statistics*
Nominal	determine equality	categorization classification	count, mode, percentage Chi-square, binomial test
Ordinal	determine equality, relative position	rank	median, Friedman two-way ANOVA, rank-order correlation, other non-parametric statistics
Interval	determine equality, relative position, magnitude of difference	index numbers, attitudes measures, perceptions	mean, standard deviation, entire range of statistics
Ratio	determine equality, relative position, magnitude of difference with a meaningful zero	sales, costs, many objective measurements	entire range of statistics

*All statistics applicable to a scale are also applicable to any higher scale in the table. For example, all the statistics applicable to nominal data can be applied to ordinal, interval and ratio data.
Modified from S.S. Steven, 'On the Science of Scales of Measurements' [28].

2.6 Scales to Measure Direct Responses

Most of the psychological scaling involved in clothing comfort research can be regarded as measurement of attitudes. Tull and Hawkins defined an attitude as an enduring organization of cognitive, affective, and behavioral components and processes with respect to some aspect of the individual's world. The three components include:

- a person's beliefs or information about the object (a cognitive component);
- a person's feelings of like or dislike concerning the object (an affective component);
- a person's action tendencies or predisposition towards the object (a behavioral component) [27].

To measure directly an individual's attitude, or a component of the attitude, attitude scales can be used on which the individual is required to explicitly state his or her attitude. Attitude scales consist of a rating scale, or a group of rating scales, that measure single dimensions of attitude components. In using rating scales, a respondent is required to place an attribute of the object being rated at some point along a numerically valued continuum or in one of several numerically ordered categories. The researcher can design rating scales to focus on different aspects, such as the overall attitude towards an object, the degree to which an object contains a particular attribute, the feeling toward an attribute, and the importance attached to an attribute.

There are two major types of rating scales: non-comparative and comparative. With a non-comparative rating scale, the respondent is not provided with a standard to use in assigning the rating. Non-comparative rating scales have two major forms: graphic rating scales and itemized non-comparative rating scales. The later are most frequently used and are the basic building blocks for the more complex attitude scales. In using itemized rating scales, the respondent is required to select one category from a limited number that are ordered in terms of their scale positions.

In developing methodology for studying human perception of clothing, Hollies [13,15] used a number of itemized rating scales for the sensations derived from participants; these are shown in Tables 2-2 and 2-3.

Table 2-2
Hollies' Four-Point Scale [15]

4	=	partially
3	=	mildly
2	=	definitely
1	=	totally

Table 2-3
Hollies' Five-Point Scale [13]

1	totally uncomfortable
2	
3	
4	
5	completely comfortable

In studying thermal comfort, McGinnis of the Army Natick Laboratories devised an intensity scale with 13 points, as shown in Table 2-4. This scale appears to be in the form of an interval scale, although it is doubtful whether the interval between each of these points is exactly equal. However, it seems that the data from such scales can be treated as if they were of equal interval, as the results of most standard statistical techniques are not affected greatly by small deviations from the interval requirements [29,30]. If there is a concern, most ordinal data can be transformed into workable interval data [31].

Table 2-4
McGinnis' Thermal Scale [15]

I am:

So cold I am helpless
Numb with cold
Very cold
Cold
Uncomfortably cold
Cool but fairly comfortable
Comfortable
Warm but fairly comfortable
Uncomfortably warm
Hot
Very hot
Almost as hot as I can stand
So hot I am sick and nauseated

Tull and Hawkins [27] listed a number of cautions in constructing and using itemized rating scales:

- The nature and degree of verbal category descriptions affect the responses.
- The number of categories can be created in many ways, depending on the purpose and nature of the investigation. Five categories can be used when several scales are to be summed for one score. Up to nine categories can be used when interested and knowledgeable respondents are comparing attributes across objects.
- The decision to use a balanced or an unbalanced set of categories depends on the type of information desired and the assumed distribution of attitudes in the population studied. A balanced scale provides an equal number of favorable and unfavorable categories. Balanced scales should be used unless it is known that the respondent's attitudes are unbalanced.
- The use of an odd or even number of categories is a relevant issue when balanced scales are being constructed. If an odd number of scale items is used, the middle item is generally designed as a neutral point. Normally, odd number categories should be used if respondents could feel neutral.
- Use of forced versus unforced scales is another important issue. A forced scale requires the respondent to indicate an attitude on the item. In such a situation, a respondent may often mark the midpoint of scale when he or she has no attitude or knowledge on the item. If a sufficient proportion of the respondents act in such a way, utilization of the midpoint will distort measures of central tendency and variance. Therefore, unforced scales should be used unless it can be assumed that all respondents have knowledge on the item.

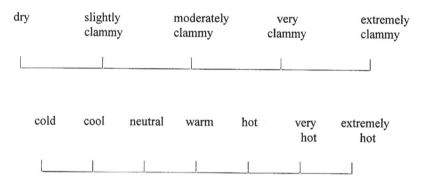

Fig. 2-1 Subjective rating scales for clamminess and thermal sensations

Hollies used unforced scales in his study of consumers' perception on clothing comfort [15], while Kawabata utilized forced scales in his study of fabric handle assessment by textile experts [21]. In studying the moisturizing buffering of hygroscopic fabrics, Li *et al.* used non-comparative unbalanced and forced rating scales, as shown in Figure 2-1. The scales were chosen on the basis of the fact that, under the testing conditions, subjects were unlikely to feel cold on the thermal scale, and the rating scale for clammy sensation could only have neutral to extreme points. Also, thermal and clammy sensations are such fundamental perceptions that subjects were sure to know the items well, especially after training [32].

In the non-comparative scales discussed above, different respondents may apply different standards or reference points when they evaluate objects without direct reference to specific standards. When asked to rate the overall comfort or handle of a garment, some respondents may compare it to their ideal garment, others to their current similar garment, and still others to their impression of the average for the same type of garment. Therefore, when we want to ensure that all respondents are approaching the rating task from the same known reference point, comparative rating scales should be used.

Non-comparative rating scales (both graphic and itemized) can be converted to comparative scales by simply introducing a comparison point. The usage of non-comparative graphic and itemized rating scales and the issues discussed can apply to comparative scales. In studying the physical mechanism of dampness perception, Plante *et al.* used a comparative unbalanced rating scale (shown in Fig. 2-2), with specific 'dry' and 'very damp' fabrics as references [33,34].

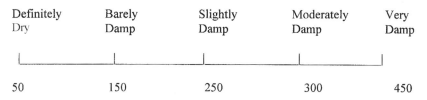

Fig. 2-2 Dampness rating scales

In determining consumers' preferences and their ability to discriminate among products, pairwise measurement methods can be used. There are various ways to conduct pairwise measurements, including paired comparisons, double-paired comparisons, consistent preference discrimination tests, triangle discrimination and triangle preference tests, and response latency. Paired comparison is most frequently used in clothing comfort research. The use of the technique involves presenting the respondents with two objects at a time and requiring the selection of one of them according to some criterion. Each respondent must compare all possible pairs of objects ($n[n-1]/2$, where n is the number of objects being studies). For each attribute of interest, a comparison needs to be conducted. Due to the large number of tests involved, paired comparisons are generally limited to one attribute, such as overall preference, or to a couple of products on multiple attributes. The outputs from paired comparisons can be analyzed in a number of ways. A simple visual inspection can reveal the preferences for one product over another, which can then be used for judging the rank order among a number of products. The data can also be converted into an interval scale through the application of Thurstone's law of comparative judgment [35].

Many researchers report use of paired comparisons in studying the comfort attributes of textile products. Fuzek and Ammons applied paired comparison techniques to obtain subjective evaluations of comfort performance of T-shirts [36]. Li *et al.* used paired comparison techniques, together with non-comparative rating scales, to obtain the overall preference of consumers towards T-shirts made from eight types of fibers, through handling and wearing experience. The preference outputs were converted into an interval scale and used to study its relationships with various sensations and physiological responses [37] and fabric physical properties [38]. Schneider *et al.* utilized a paired comparison method to study the coolness to the touch of hygroscopic fibers [39,40].

The ranking scores obtained through paired comparisons are essentially ordinal data, which are effective in obtaining the preferences of consumers when comparing a series of products. However, they cannot provide the magnitude of the perceived differences between samples, nor their relative positions in a context of all possible relevant samples beyond a particular experiment.

Rank order rating scale is another scale widely used to measure preference for comfort attributes, with which respondents are required to rank a set of objects according to some criterion. Like paired comparison, this method is purely comparative in nature, and its outputs are only applicable within the product range being studied. The rank order method forces respondents to discriminate among the relevant objects in a manner close to the actual shopping environment. It is less time-consuming than paired comparisons and the instructions for ranking are easily understood by most individuals. The major shortcoming of this technique is that it produces only ordinal data with which the number of statistical analyses permissible is limited. In Hollies' wear trials, the Wilcoxon Sign-Rank Test was used to detect significant differences in ranking [15].

On the basis of rating scales, more complex attitude scales can be constructed to measure more aspects of an individual's attitude toward some object. The responses from respondents to various scales can be summed to provide a single attitude score for the individual. More commonly, the responses to each scale item or subgroup items may be examined independently. Hollies *et al.* developed a comprehensive attitude scale (a rating sheet shown in Fig. 2-3) to obtain various sensory responses from respondents in wear trials [15]. Normally, the sensory responses of

Comfort Description	Minutes in Environmental Chamber					
	0	15	30	45	60	75
Stiff	—	—	—	—	—	—
Staticky	—	—	—	—	—	—
Sticky	—	—	—	—	—	—
Nonabsorbent	—	—	—	—	—	—
Cold	—	—	—	—	—	—
Clammy	—	—	—	—	—	—
Damp	—	—	—	—	—	—
Clingy	—	—	—	—	—	—
Picky	—	—	—	—	—	—
Rough	—	—	—	—	—	—
Scratchy	—	—	—	—	—	—
McGinnis Scale	—	—	—	—	—	—

Comfort Intensity Scale

1 — 2 — 3 — 4 — 5

Totally   Completely
Uncomfortable   Comfortable

Fig. 2-3 Hollies' subjective comfort rating sheet [13]

subjects were analyzed individually. Li *et al.* applied similar multi-attitude scales to study the comfort performance of sportswear made from different fibers. The responses to various sensory items were first analyzed individually [37], then the relationships among the sensory responses were investigated by factor analysis and clustering analysis [41].

In studying fabric hand properties, both Elder *et al.* [42] and Mackay [43] applied magnitude-estimation technique. The technique requires subjects to estimate the magnitude of an attribute of fabrics by comparing them with a standard fabric or with their own experience. Their estimations are recorded by assigning a number, or marking a position on a line for each fabric sample. The scales are open. Subjects can use scores that seem appropriate to them. Magnitude-estimation technique was also used by Sweeney and Branson in studying the psychophysical mechanism of dampness perception [44].

Another frequently used attitude scale in sensory research is the semantic differential scale. Semantic differential scales were developed by Osgood *et al.* in studying the meaning of language [45]. Kelly developed a similar technique using a 'repertory grid' [46]. Semantic differential scales consist of a series of bipolar rating scales, each of which is made of word pairs that may be opposites or one extreme and one neutral pole. The bipolar words are bounded on a number of itemized, five to seven rating scales at each end. In the case of opposites, the center is neutral between the two extremes. Respondents are instructed to mark the blank that best indicates how accurately one or other term describes or fits the attitude object.

Semantic differential rating scales can have any number of scale points, with six to seven being most common. Friedman *et al.* recommended that the more favorable adjective or phrase be randomly assigned to the left and right side of the scale [47]. Due to widespread use, semantic differential scales have been improved in many forms, such as the upgraded semantic differential (or graphic position) scale [48], the numerical comparative scale [49], and the Stapel scale (a simplified version of the semantic differential scale) [27].

A widely accepted assumption is that the resultant data from semantic differential scales are interval in nature. There are a number of ways to analyze semantic differential data. Two approaches, aggregate analysis and profile analysis, are widely used. For aggregate analysis, the larger numbers are consistently assigned to the blanks nearer the more favorable terms, and the scores across all adjective pairs are summed for each individual. The individual or group of individuals can be compared to other individuals on the basis of total scores. Different objects can be also compared for the same group of individuals. Aggregate analysis is most effective for predicting overall preferences. In profile analysis, the mean or median is calculated for each adjective pair for an object by a specified group of respondents. Then, this profile can be compared with the profiles of other objects, other groups, or the 'ideal' version of the object. Profile analysis is useful to isolate strong attributes of products [27].

A number of researchers used semantic differential scales in clothing comfort studies. Winakor *et al.* [50] and Chen *et al.* [51] used 99-point scales with bipolar pairs to study fabric sensory properties. During judgment, a participant is asked to assign a number between 1 and 99 with 1 = complete agreement with the left-hand descriptor, 99 = complete agreement with the right hand descriptor, and 50 = uncertain. The scores are transformed in considering the non-uniform distribution of sensitivity along the scale.

Fritz [24,52] applied semantic differential scales to study the handle of fabrics. Seven point scales were used with 3 at the two ends indicating 'extremely', 0 at the middle point indicating 'neither both'. Profile analysis was applied to compare ideal underslip fabrics with the perceptions of male and female respondents. Fig. 2-4 shows an example of her work.

Byrne *et al.* [53] applied semantic differential scales to investigate the sensory perceptions of consumers on fiber types for different end-uses. Profile analysis was applied to compare the responses from Australian and English students. Aggregate analysis was used to sum the differences between the ideal and each fiber type as a measure of the extent to which the fiber deviates from the ideal.

Bishop pointed out that, in the context of fabric objective measurement, bipolar descriptors do not have any added value over single descriptors, but have a number of disadvantages. Bipolar descriptors may complicate the process of descriptor generation, impose unnecessary correlation between descriptors, represent positive and negative fabric attributes in the minds of respondents, and introduce an element of liking/disliking in an assessment [54].

	Extremely	very much	some what	neither both	some what	very much	extremely	
Soft	3	2	1	0	1	2	3	Harsh
Smooth	3	2	1	0	1	2	3	Rough
Cool	3	2	1	0	1	2	3	Hot
Light	3	2	1	0	1	2	3	Heavy
Fine	3	2	1	0	1	2	3	Coarse
Crisp	3	2	1	0	1	2	3	Limp
Clammy	3	2	1	0	1	2	3	Absorbent
Natural	3	2	1	0	1	2	3	Synthetic
Sheer	3	2	1	0	1	2	3	Bulky
Clingy	3	2	1	0	1	2	3	Flowing
Crushable	3	2	1	0	1	2	3	Resilient
Lacy	3	2	1	0	1	2	3	Plain
Drapable	3	2	1	0	1	2	3	Rigid
Scratchy	3	2	1	0	1	2	3	Silky
Stiff	3	2	1	0	1	2	3	Soft

Fig. 2-4 Fritz's semantic differential scale [24]

From these discussions, we can see that various attitude scaling techniques have been applied to measure sensory perception of clothing comfort. Besides these scales, there are numerous less well-known methods, and modified versions of the popular scales. In comparing various scaling methods, the results generally have been equivalent across the techniques [55]. Selection of a scale technique depends upon a number of issues: the required information, the characteristics of the respondents, the means of administration, and the cost involved. Generally speaking, multiple measures are more effective than any single technique. The sum of several items gives a more accurate measurement than a single measurement [27].

2.7 Wear Trial Technique

Perceptions of sensory comfort of clothing may involve various sensory channels from all the five senses: visual, auditory, smell, taste, and touch, but are mainly associated with skin sensory systems. Many comfort sensations can only be generated under certain wear situations with the existence of relevant physical stimuli. The physical stimuli such as heat, moisture, and mechanical stimulation from fabric to the skin often can be generated only under specific combinations of physiological states (e.g. sweating rate), fabric materials, garment fitness, and environmental conditions (e.g. temperature, humidity, and air velocity). A large amount of research work has been published on fabric sensory properties through hand [54]. Human beings often use hands to obtain tactile information as well as to manipulate objects. However, Stevens reported that much of the tactile sensations come from parts of the body other than the hands [56]. This suggests that perception of comfort performance of clothing has to be studied in wear situations. Therefore, wear trialing is an important technique for clothing comfort research.

On the basis of their research work on human perception and clothing comfort, Hollies *et al.* developed a wear trial experimental technique to characterize the sensory comfort of clothing. The technique included a number of components: (i) generating sensory descriptors with respondents; (ii) selecting testing conditions to maximize the opportunities for perception of various sensations; (iii) designing attitude scales in the way of rating sheets to obtain various sensory responses to particular garments (see Fig. 2-3); (iv) conducting wear trials in controlled environmental chambers according to predetermined protocol; (v) collecting data and analyzing and interpreting the results. Table 2-5 shows the typical microclimate and exercise protocol used in the work. This technique has been extensively applied by Hollies and his colleagues to evaluate various apparel products [13,15–17].

Table 2-5
Microclimate and Exercise Protocol used in Hollies' Wear Trials [15]

Rating period	Time in chamber (min)	Exercise before rating[a]	Air temper -ature (°C)	Relative humidity (%)
1	0	no	35	20
2	1	yes	35	20
3	15	no	35	70
4	16	yes	35	70
5	30	no	35	70
6	31	yes	35	70
7	45	no	35	45
8	60	no	21	75
9	75	no	17	75

[a]Exercise time, 20 seconds
Note: Subjects exercise in antechamber for 10 min at 150 to 180 kg cal/m^2 hour at 30°C to 33°C

Li *et al.* adopted the principal of this technique and designed a multi-attitude scale and wear trial protocol to study physiological responses, sensory perceptions, and preferences of consumers towards sportswear made from eight types of fibers [37]. The trial protocol is shown in Table 2-6. Subjective preferences and sensory responses were obtained by exposing subjects to periods of exercise (30 min) and rest, in two environmental conditions: hot (32°C, 45% r.h.) and cold (14°C, 32% r.h.). The air velocity was 0.25 m/s. Every five minutes during exercise, measurements were made of tympanic membrane and skin temperatures, heart rate, and energy expenditure. Body sweat loss and sweat absorption of the garments were also recorded. Every 10 minutes, subjective responses to 19 sensory descriptors were recorded on a scale from 1 (no sensation) to 5 (totally).

The sensory descriptors included *snug, loose, heavy, lightweight, soft, stiff, staticky, sticky, nonabsorbent, cold, clammy, damp, hot, clingy, sultry, prickly, rough, scratchy,* and *itchy.* Overall preferences between the two garments tested in each trial were recorded after handling at the beginning of the trial, and again after wear at the end of the experiment. This trial applied multiple scaling techniques. The overall preferences through wearing and handling were obtained by paired comparison design, while subjective sensory responses were obtained through a comprehensive attitude scale, similar to Hollies' rating sheet shown in Fig. 2-3.

Table 2-6
Protocol for the Psycho-physiological Trials [37]

Time (min)	Subjective rating period	Physiological measurement period	Climate in the chamber				Note
			hot		cold		
			T (°C)	r.h. (%)	T (°C)	r.h. (%)	
0	✓		20	50	20	50	
10	✓		20	50	20	50	
Evaluate the T-shirts by handling and give initial preference between the two							
0	✓	✓	32	45	14	32	wearing the 1st
5	✓	✓	32	45	14	32	T-shirt and doing
10	✓	✓	32	45	14	32	exercise
15	✓	✓	32	45	14	32	
20	✓	✓	32	45	14	32	
25	✓	✓	32	45	14	32	
30	✓	✓	32	45	14	32	
0 20	Rest in room temperature (20°C and 50% r.h.), then change into the second T-shirt						
0	✓	✓	32	45	14	32	wearing the 2nd
5	✓	✓	32	45	14	32	T-shirt and doing
10	✓	✓	32	45	14	32	exercise
15	✓	✓	32	45	14	32	
20	✓	✓	32	45	14	32	
25	✓	✓	32	45	14	32	
30	✓	✓	32	45	14	32	
0 5	Evaluate the overall comfort performance of the two T-shirts, give overall preference between the T-shirts						

Researchers in CSIRO Division of Wool Technology adopted Hollies' technique, and set up two climatic chambers to study the comfort performance and attributes of wool apparel products. A large number of wear trials were conducted in the chambers to study the comfort features of wool garments under various wear situations. For instances, the moisture buffering effect of clothing during exercise was reported in 1992 [32], the thermoregulatory responses of the body during intensive exercise when wearing hygroscopic fibers in 1993 [57], the perception of dampness in 1995 [33], and the perception of fabric coolness in 1996 [40].

2.8 Dimensions of Comfort Sensory Perceptions
Data analysis is extremely important in comfort sensory study. Traditionally, it has been performed with univariate statistical tools such as Analysis of Variance and Correlation Analysis. In the last decade, multivariate techniques have become available for increasing the understanding of complex data because of greatly increased computing capacities. The development in this area has been very fast. The common principle of various techniques is to extract central or common information in large data volumes and present them in understandable and simplified forms.

Stein and Meredith argued from the view of physiology of perception that, although some modality-specific characteristics may be largely preserved as the brain sorts out the inputs from many cues, others are certainly altered. The brain does not perceive the world as a series of independent sensory experiences; rather there is an interweaving of different sensory impressions through which sensory components are subtly altered by, and integrated with, one another. The product of these integrative processes is the perception [58].

Risvik (1996) pointed out that human minds never perceive a product as a sum of attributes. Our minds may focus on key attributes, aggregate attributes into concepts, perceive holistic forms, or make up an iterative process with mixtures of the aforementioned for consultation when some decisions are to be made. When a product is perceived, our brain may contain fewer attributes/concepts than would be required for a complete sensory profile, which implies that some form of aggregation of information is taking place in our information processing [59].

Risvik further suggested that words for sensory profiling have a number of features:

- Words describing product attributes have different levels of complexity that cannot be defined.
- Complex words may be related to several aspects of product perception, which can be defined as a fuzzy latent structure, on the basis of the nature of the words and utilization of the language.
- In a sensory profiling of a product, it is not unusual to have 15–20 words describing its attributes, most of which are interacting and overlapping.

The majority of the words show only slightly different perspectives on the understanding of the product. Therefore, it is of the utmost importance in performing data analysis to be able to handle these problems and to utilize the information for interpretation in the analysis.

In studying fabric softness and stiffness, Elder *et al.* [42,60] found that subjective perceptions of fabric attributes, such as stiffness and softness, may be derived from a combination of fabric physical properties rather from a single physical property. This combination may vary with fabric construction and end-uses.

In 1956, Miller claimed that humans could only handle 5-9 independent phenomena simultaneously in their consciousness [61]. On the other hand, Martens pointed out that, in the analysis of sensory profiling data, most often 1-3 dimensions contain the essential information in the data [62]. These contribute to an assumption that the complex human sensory perceptions can be reduced to a few independent dimensions in human consciousness, called latent variables or latent phenomena. In statistical terms, latent variables are the projections or linear combinations of several variables in the data. This implies that statistical analysis may reflect the process of human perception of a product. Each perceived dimension is a combination of contributions from various product attributes [59].

From these perspectives, multivariate statistical methods have been widely used to study the perception of products in the food industry, and to describe consumers' perceptions of brands in marketing research. These multivariate statistical analyses have a number of objectives:

- to identify the number of dimensions that respondents use to distinguish different products,
- to reveal the nature or characteristics of these dimensions,
- to locate products on these dimensions as consumers perceive them, and
- to determine the ideal or preferred location of a product on each of the dimensions.

There are a number of statistical tools that can be used to achieve these objectives, including clustering analysis, principal component analysis (PCA), factor analysis, discriminate analysis, and correspondence analysis. In 1968, Yoshida conducted a series of experiments designed to discover the dimensions of tactual impressions, by using stimulus samples differing in size, shape, and texture. Through factor analysis, he found that 70% of the variance could be accounted for by three dimensions: (i) heaviness–coldness, (ii) wetness–smoothness, and (iii) hardness [63].

Using the oblique principal component cluster analysis method to analyse the sensory responses discussed in Section 2.3, Li *et al.* [22] obtained the results shown in Table 2-7, in which the variables in each cluster are listed. Two squared correlation coefficients are listed for each cluster.

The column labeled R^2 – *Own Cluster* gives the squared correlation with its own cluster component. The larger the squared correlation is, the closer the association is. The column labeled R^2–*Next Closest* contains the next highest squared correlation of the variable with a cluster component. This value is low if the clusters are well separated. The column headed *1–R^2 Ratio* gives the ratio of one minus the R^2 – *Own Cluster* to one minus the R^2 – *Next Closest*. A small *1–R^2 Ratio* indicates a good clustering.

By comparing the results from Table 2-7 with those for summer wear and sportswear, the general pattern of the cluster analysis showed that the 26 sensory descriptors could be classified into four clusters. The basic components of the four clusters are:

- Cluster 1 – tactile sensations: *prickly, tickling, rough, raggy, scratchy, itchy, picky, staticky;*
- Cluster 2 – moisture sensations: *clammy, damp, wet, sticky, sultry, nonabsorbent, clingy;*
- Cluster 3 – pressure (body-fit) sensations: *snug, loose, lightweight, heavy, soft, stiff;*
- Cluster 4 – thermal sensations: *cold, chill, cool, warm,* and *hot.*

Cluster 1, of tactile sensations, was the most stable group, in which the components of the cluster were relatively well defined and did not change much with the types of clothing. However,

Table 2-7
Cluster Pattern between the 26 Sensory Descriptors for Winter Wear

Cluster	Variable	R^2 – Own Cluster	R^2 – Next Closest	1–R^2 Ratio
Cluster 1	Prickly	0.65	0.21	0.44
	Tickling	0.62	0.23	0.50
	Rough	0.59	0.19	0.50
	Raggy	0.52	0.13	0.55
	Scratchy	0.44	0.15	0.65
	Itchy	0.44	0.16	0.66
	Picky	0.42	0.10	0.65
	Heavy	0.20	0.06	0.85
Cluster 2	Clammy	0.53	0.17	0.56
	Sticky	0.52	0.12	0.54
	Sultry	0.42	0.10	0.64
	Nonabsorbent	0.40	0.06	0.64
	Damp	0.39	0.08	0.66
	Clingy	0.36	0.12	0.73
Cluster 3	Hot	0.51	0.10	0.54
	Soft	0.47	0.01	0.54
	Snug	0.36	0.05	0.67
	Warm	0.32	0.01	0.68
	Loose	0.26	0.05	0.78
	Lightweight	0.25	0.04	0.79
Cluster 4	Cold	0.54	0.16	0.54
	Chill	0.53	0.18	0.57
	Wet	0.36	0.25	0.85
	Stiff	0.30	0.13	0.80
	Staticky	0.29	0.12	0.81
	Cool	0.27	0.06	0.78

some sensations from other clusters such as *heavy, stiff, clingy,* and *clammy* joined in when they became more closely associated with this cluster in certain wear conditions. Cluster 2, of moisture sensations, was also relatively stable. Its components were relatively well clustered and did not change much with the types of clothing. However, it showed interaction with thermal sensations *hot* and *chill* in sportswear, and interacted with tactile sensations in summer wear. Cluster 3 of pressure sensations and Cluster 4 of thermal sensations were not stable. Their components were not clearly clustered and changed their membership frequently. The pressure sensations can be synthetic sensations, showing interaction with tactile and thermal sensations.

The perception of thermal sensations is heavily dependent on wear situations, and interacts strongly with moisture sensations. For instance, the sensation *hot* is associated with the pressure cluster for winter wear, but with the moisture cluster for summer wear and sportswear, which is a reasonable change of association. From our daily experience, the sensation *hot* is frequently perceived with *snug* and *soft* clothing in winter, but is perceived more frequently together with the moisture sensations *damp, wet, sultry,* and *sticky* in summer and sporting wear situations.

By non-parametric clustering analysis of the data reported by Hollies *et al.* over the years [13,15–17], the authors found that the sensory descriptors used by them could be grouped into two clusters: Cluster 1 (*scratchy, rough, picky, stiff, heavy, lightweight, loose,* and *snug*) and Cluster 2 (*damp, clammy, sticky, clingy,* and *nonabsorbent*). These two clusters mainly correspond with the tactile and moisture clusters.

Li also applied factor analysis and clustering analysis to the sensory data obtained from wear trials under controlled physical activity and environmental condition (32°C, 45% r.h.) [65]. Table 2-8 shows the results from factor analysis on the sensory data obtained in this hot environment. Three factors whose eigenroots (sums of the squares of the factor loading) exceeded 0.89 were extracted. Each factor is interpreted by studying the factor loading of each column. The values of factor loadings between 0.200 and –0.200 were regarded as insignificant.

By comparing the results from both factor analysis and variable clustering analysis, it was found that the 19 sensory descriptors could be grouped into three factors:

Table 2-8
Varimax Rotated Factor Matrix

	Factor 1	Factor 2	Factor 3
Sultry	0.808	–	–
Damp	0.798	–	–
Sticky	0.778	–	–
Hot	0.740	0.206	–
Clingy	0.729	–	–
Nonabsorbent	0.567	–	–
Clammy	0.529	–	–
Prickly	–	0.872	–
Itchy	–	0.868	–0.244
Scratchy	–	0.841	–
Rough	–	0.816	–
Lightweight	–	–	0.749
Soft	–	–0.215	0.693
Loose	–0.224	–	0.517
Snug	–	–	0.479
Staticky	0.282	0.283	–
Cold	–	–	
Heavy	0.261	–	–0.302
Stiff	–	0.260	–0.293

- Factor 1 – thermal and moisture sensations: *sultry, damp, clammy, clingy, hot, cold,* and *nonabsorbent*;
- Factor 2 – tactile sensations: *prickly, scratchy, rough, itchy,* and *staticky*;
- Factor 3 – pressure sensations: *snug, loose, heavy, lightweight, soft,* and *stiff*.

Similar patterns of relationships among the psychological responses to the 19 sensory descriptors were found in another set of sensory data obtained in a cold environment (14°C, 32% r.h.) [41]. These analyses illustrated a pattern of relationships among the sensory responses, which were rather vague but which were stable and consistent with different statistical methods, with different wear conditions (hot and cold), and with the psychological perceptual data collected from a large number of independent studies. From these results, we can conceptualize that the comfort of clothing has three latent independent sensory factors (dimensions): *thermal-wet comfort, tactile comfort, and pressure comfort.*

Thermal-wet comfort is mainly related to the sensations involving temperature and moisture, such as *sultry, clingy, hot, damp, clammy, cold, nonabsorbent,* and *sticky*. This factor responds mainly with the thermal receptors in skin and relates to the transport properties of clothing such as heat transfer, moisture transfer, and air permeability (see Sections 3 and 6).

Tactile comfort is associated with the sensations involving direct skin–fabric mechanical interactions, such as *prickly, scratchy, itchy, rough,* and *staticky*. This factor responds largely with the pain receptors in the skin and relates mainly to the surface characteristics of the fabric, including the diameter of fiber ends, its density, and the smoothness of its surface (see Sections 3 and 7).

Pressure comfort is more complex and involves a number of synthetic sensations such as *snug, loose, heavy, lightweight, soft,* and *stiff*. This factor may mainly correspond to the pressure receptors in skin and may come from some combination of a number of simple sensory responses. Fabric bulk, mechanical behavior, and overall fit of garment to the body may be responsible to this dimension of comfort. Fabric handle properties are also highly related to this factor (see Sections 3 and 7).

In 1993, Byrne *et al.* conducted a consumer perceptual study on fibers types and end-uses, in the United Kingdom and Australia [53]. The sensory descriptors were generated by consumers in the form of semantic differential grids. The fibers studied were silk, cotton, polyester, and nylon. The end-uses were sportshirts and underslips. Principal Component Analysis was applied to the sensory data collected. They found two factors accounted for 90% of the total variance in the data for sportshirts. The trials contributing to the two factors were:

Component 1   **Component 2**
synthetic–natural   *flimsy–substantial*
comfortable–itchy   *limp–crisp*
harsh–soft   *thin–thick*
cool–clammy
clammy–fresh
dense–loose
sweaty–absorbent.

They also applied principal analysis to the sensory data obtained for underslips. Three factors accounted for 94% of the variance were found, namely:

Component 1	**Component 2**	**Component 3**
soft–harsh	*clammy–absorbent*	*scratchy–silky*
smooth–roughness	*cool–hot*	
light–heavy	*crisp–limp*	
fine–coarse	*natural–synthetic*	
soft–stiff		
sheer–bulky		
drapable–rigid		
clingy–flowing		

Comparing these results with those reported by Li *et al.*, Component 1 appears to correspond to the *pressure* comfort factor, Component 2 to the *thermal-wet* comfort factor, and Component 3 to the *tactile* comfort factor. This further confirms the observation that there are three major latent dimensions in clothing comfort sensory perceptions. From these findings, it seems reasonable to accept the assumption that the complex sensory human perceptions on clothing comfort can be reduced to around three independent dimensions or latent variables.

2.9 Overall Comfort Perception and Preferences

Understanding how consumers perceive clothing and formulate their preferences is of compelling interest to both researchers and manufacturers. The overall sensory perception and preferences of a wearer to the clothing he or she wears are the result of a complex combination of sensory factors that come from the integration of inputs from various individual sensory modalities such as thermal, pressure and pain sensations. The individual sensory modalities are related to different mechanical–physical attributes of the garments. The sensory perceptions are also influenced by the psychological and physiological state of the individual wearers and the external environment.

The process of integration is critical for developing an understanding of the psychological picture of clothing comfort. Subjective preference is further integration from inputs of the integrated sensory impressions in reference with past experiences, psychological desires, and the physiological status of the wearer, to form a final assessment of clothing. The integrated sensory impressions are highly related to the sensory factors that are derived from the latent pattern in various sensations. The relative contributions of the sensory factors to subjective preference may be different under different wear situations, since the psychological and physiological requirements of a wearer to clothing are dependent on specific combinations of the physical activities of the individual and the external environmental conditions.

Li [64] investigated the inter-relationships and predictability between the sensory factors and the subjective preference votes, by canonical correlation analysis. The study was based on measurements of sensory responses and subjective preference votes to eight kinds of knitted T-shirts in a psycho-physiological wearer trial, and the three sensory factors derived as discussed in Section 2.8.

Canonical correlation is a technique for analyzing the relationships between two sets of variables. Each set can contain several variables. A statistical program (CANCORR), from the SAS software package, was used to perform the analysis. Given two sets of variables, CANCORR finds a linear combination from each set, called a canonical variable, such that the correlation is maximized. This correlation between the two canonical variables is the first canonical correlation.

The coefficients of the linear combinations are canonical coefficients or canonical weights. CANCORR continues by finding a second set of canonical variables, unrelated to the first pair, which produces the second highest correlation coefficient. The process of constructing canonical variables continues until the number of pairs of canonical variables equals the number of variables in the smaller group. Each canonical variable is not correlated with all the other canonical variables of either set, except for the one corresponding canonical variable in the opposite set [67–69].

Canonical redundancy analysis [67] is a technique that examines how well the original variables can be predicted from the canonical variables. Redundancy is the proportion of variance extracted by a canonical variable multiplied by the proportion of shared variance between the canonical variable and the corresponding canonical variable of the other set. The square of the canonical correlation coefficient is a measure of the overlap between the two canonical variables. The total redundancy of the canonical variables for the original variables is the sum of the redundancy of the individual variable and is called 'cumulative proportion'.

Table 2-9
Canonical Correlation Analysis between the Subjective Preference Votes and the Psychological Sensory Factors

Type*	Name*	FACT1	FACT2	FACT3	HMVOTE	LMVOTE
CORR	FACT1	1.000	0.132	−0.272	−0.068	0.122
CORR	FACT2	0.132	1.000	−0.525	−0.893	−0.450
CORR	FACT3	−0.272	−0.525	1.000	0.730	0.741
CORR	HMVOTE	−0.068	−0.893	0.730	1.000	0.569
CORR	LMVOTE	−0.122	−0.450	0.741	0.569	1.000
CANCORR		0.969[a]	0.580[b]	−	−	−
SCORE	PSYC1	0.211	−0.614	0.561	0.000	0.000
SCORE	PSYC2	0.506	0.932	1.008	0.000	0.000
SCORE	VOTE1	0.000	0.000	0.000	0.817	0.275
SCORE	VOTE2	0.000	0.000	0.000	−0.901	1.185
STRUCTUR	PSYC1	−0.023	−0.881	0.826	0.944	0.718
STRUCTUR	PSYC2	0.355	0.469	0.380	−0.131	0.390
STRUCTUR	VOTE1	−0.022	−0.854	0.801	0.974	0.741
STRUCTUR	VOTE2	0.205	0.272	0.221	−0.226	0.672

Note:
[a] $P < 0.0001$
[b] $P < 0.01$
*COR indicates that the numerical values in the row are Pearson correlation coefficients.
CANCORR indicates that the numerical values in the row are canonical correlations.
SCORE indicates that the numerical values in the row are the standardized canonical coefficients.
STRUCTUR indicates that the numerical values in the row are the correlation between each of the canonical variables and the factors or preference votes.
PSYC1 and PSYC2 are the canonical variables from the psychological sensory factors.
VOTE1 and VOTE2 are the canonical variables from the preference votes.
HMVOTE is the preference vote after wearing.
LMVOTE is the preference vote by handling.
The other abbreviations are defined in the text.

The squared multiple correlation is the sum of squared correlation coefficients of each factor with the first *m* canonical variables of the opposite set, where *m* varies from 1 to the number of canonical correlation coefficients. These squared multiple correlation coefficients indicate the predictive power of the canonical variables for each of the original variables of the other set. The results of canonical correlation analysis for the psychological sensory factors and the subjective preference votes are summarized in Tables 2-9, 2-10, and 2-11.

Two significant canonical correlation coefficients were found, suggesting that there are two independent sensory factors of the psychological sensations that are significantly related to the subjective preference votes.

The first canonical correlation (0.969) indicated that the overall preference votes after wearing were very closely related to *tactile* and *pressure* comfort factors. Spearman correlation analysis was applied to the psychological responses to the 19 sensory descriptors obtained from the wearer trials and the subjective preference votes. It was found that the overall preference votes after wearing were significantly correlated only with the sensory descriptors *prickly, rough, scratchy, itchy, heavy, stiff,* and *soft.*

The second canonical correlation (0.580) suggests that the preference votes by handling are mainly related to the *pressure* comfort factor. It was found that the preference votes by handling were significantly correlated with the sensory descriptors *heavy, stiff,* and *lightweight* in the Spearman correlation analysis.

The canonical redundancy analysis showed that the canonical variables of sensory factors are reasonably good predictors of the subjective vote canonical variables, with a cumulative redundancy of 0.788. The canonical variables of the subjective preference votes, on the other hand, are not such good predictors for the sensory factor canonical variables, with a cumulative redundancy of 0.512.

This suggests that sensory factors can predict subjective preference votes quite well, but not *vice versa*, indicating that consumers make their preference judgments largely on the basis of their sensory perceptions.

Table 2-10
Canonical Redundancy Analysis

Standardized variance of the sensory factors explained by:

their own canonical variables			the opposite canonical variables	
Proportion	Cumulative proportion	Canonical R-squared	Proportion	Cumulative proportion
1 0.487	0.487	0.939	0.457	0.457
2 0.164	0.650	0.336	0.055	0.512

Standardized variance of the subjective votes explained by:

their own canonical variables			the opposite canonical variables	
Proportion	Cumulative proportion	Canonical R-squared	Proportion	Cumulative proportion
1 0.749	0.749	0.939	0.703	0.703
2 0.251	1.000	0.336	0.085	0.788

This interpretation is confirmed by the squared multiple correlation coefficients, which indicate that the first two canonical variables for the sensory factors have very good predictive power for the overall preference votes after wearing, but only moderate power for the handling votes. On the other hand, the first two canonical variables for the subjective preference votes have good predictive power for the *tactile* comfort factor, fairly good predictive power for the *pressure* comfort factor, but are almost useless for the *thermal-wet* comfort factor. These results indicate that the preference judgment by handling is different from the preference judgment by wearing. During the handling process, only a part of the sensory features of the clothing can be perceived through the hands, and this is mainly related to the *pressure* sensations.

Table 2-11
Squared Multiple Correlation Coefficients s between the First *m* Canonical
Variables and the Opposite Set of Variables

VOTE	FIRST 1	FIRST 2	PSYC	FIRST1	FIRST 2
FACT1	0.001	0.043	HMVOTE	0.891	0.909
FACT2	0.729	0.803			
FACT3	0.641	0.690	LMVOTE	0.515	0.667

Note: Definitions as in Table 2-9.

In an attempt to test the validity of these findings, the same approach was applied to the sensory data obtained from another series of wear trials, which were conducted under cold environmental conditions (14°C and 32% r.h.), using similar garments [66]. The results from canonical correlation analysis of the data from the cold environment agreed well with those from the hot environment. Two canonical correlation coefficients were obtained: 0.973 and 0.561. The first indicated the close relationship between the *tactile* comfort factor and preference vote by wearing, and the second between the *pressure* comfort factor and preference vote by handling. Also, canonical redundancy analysis showed that the canonical variables had relatively good predictive power for the subjective preference votes. The reverse was not true. Again, the squared multiple correlation coefficients showed good relationships of the preference votes with the *tactile* and *pressure* comfort factors, but not with the *thermal-wet* comfort factor. This suggests that, under both hot and cold environmental conditions, consumers make their preferences based on the same integrated sensory impressions (factors) for next-to-skin garments.

In these studies, the overall preferences were obtained on the basis of the same style and color in the garments, as the investigation was focusing on the sensory perceptions. On the other hand, this is the limitation of the studies. The components of aesthetic comfort, cost, and store or wear atmosphere were absent when the respondents were making their judgement. Therefore, the outcomes of the study cannot be directly applied to the processes of purchase decision by consumers; the conclusions can only be applied to the test conditions or similar wear situations. For different wear conditions, the relative contributions of different sensory factors change [252].

3. COMFORT AND NEUROPHYSIOLOGY

3.1 Senses and Comfort

As discussed in Section 1, clothing functions as the interface between the body and the environment. It interacts with both and influences the comfort perceptions of the wearer. In daily wear, the body, clothing, and environment all contribute to the comfort status and satisfaction of the wearer. These three elements together provide multi-sensory experiences which consist of all the information through the senses: sight, touch, kinesthetics, hearing, taste, and smell.

Obviously, visual perception is probably the most important factor influencing aesthetic comfort of clothing. Kinesthetics is the perception of body movement through the nerve endings that register the stretch or contraction of the muscle. Clothing can influence kinesthetic perception by providing restriction and pressure to muscle movement. For instance, tight-fitting Lycra cycling pants can enhance the kinesthetic perception in the muscles of the wearer's legs. Smell can be an important factor in comfort related to the body. Favorable smell from clothing may enhance the comfort perception of the wearer, while unfavorable smell may cause a feeling of discomfort. Occasionally, the sound generated from clothing influences the comfort sensory perception of the wearer. For instance, the sound of electric discharge during undressing a synthetic garment can enhance the discomfort perception caused by the electric insults on the skin. Probably, taste is the least important factor influencing clothing comfort.

Covering most parts of our body for most of the time in our daily life, clothing contacts most parts of the skin dynamically and frequently. This produces various mechanical, thermal, chemical, or electrical stimuli. Therefore, the major contributor to our sensory comfort in clothing is touch, which can be defined as the variety of sensations evoked by the stimulation from various external stimuli to the skin. The sensations perceived from these stimuli influence our overall state of comfort. The types of sensations perceived from clothes depend on how the fabric interacts with the skin and which sensory receptors are triggered. To obtain a clear understanding of these processes, we need to know the mechanisms of how our skin sensory system works. This is the focus of this section.

3.2 The Neurophysiological Basis of Sensory Perceptions
3.2.1 Skin Stimuli and the Skin Sensory System
Human skin has a very complex structure as may be seen from Fig. 3-1, which shows the structures in hairy skin covering most of the human body. The skin has two layers: the epidermis and the dermis. The epidermis is the outer layer, consisting of several layers of dead cells on top of a single living cell. The dermis is the inner layer, containing most of the nerve endings in the skin. In addition, sweat glands, hair follicles, and fine muscle filaments are housed here. Below the dermis there are layers of connective tissue and fat cells [71,72].

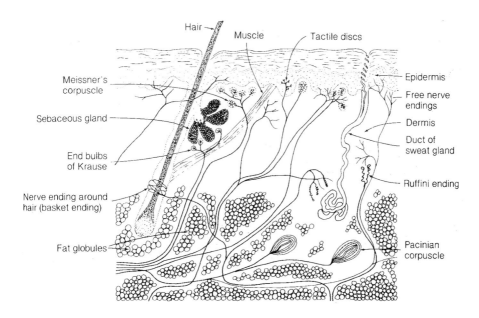

Fig. 3-1 Schematic section of human skin [72]

Fig. 3-1 also illustrates some of the nerve endings in the skin. There are two types of nerve endings: corpuscular endings and noncorpuscular (or free nerve) endings. Corpuscular nerve endings have small bodies or swellings on the dendrites, including the Pacinian corpuscles, Meisner corpuscles, Merkle disks, and Ruffini endings, which are particularly responsive to touch stimuli. The free nerve endings in subcutaneous fat are associated with pain fibers, and those projecting into the epidermis may be associated with cold fibers or pain fibers [72].

3.2.2 Transduction

The sensory receptors have a fundamental function to transduce various external stimuli into the standard code by which nervous systems work. The central question is how the sensory receptors convert the stimuli into nerve action potentials. It has been found that the common feature of the transduction is the generation of current flows within the receptor, recorded as a potential change that is proportional to the intensity of the applied stimulus. The current flow sets up nerve action potentials at a spike initiation site, which then travel centrally along the afferent nerve fiber [73].

3.2.3 Sensory Receptors

3.2.3.1 Skin Receptors Human skin is the interface between the human body and its environment; it is richly innervated and contains specialized sensory receptors to detect various external stimuli. There are three major stimuli: (i) mechanical contacts with external objects, (ii) temperature changes due to heat flow to or from the body surface, and (iii) damaging traumatic and chemical insults. In responding to these stimuli, the skin receptors produce the sensations of touch, warmth or cold, and pain [73].

3.2.3.2 Mechanoreceptors There are two groups of mechanoreceptors: (i) encapsulated receptors, including Pacinian corpuscles, Meissner corpuscles, Krause endings and Ruffini endings, which are all innervated by fast-conducting myelinated fibers; and (ii) receptors having an organized and distinctive morphology such as the hair follicle receptors and Merkle discs. Each mechanoreceptor has a distinctive range of properties that enable it to receive and respond to a particular parameter of a mechanical stimulus. The Pacinian corpuscles detect and respond to high frequencies of displacement up to about 1500 Hz, the Meissner corpuscles and the hair follicles to middle range frequencies (20–200 Hz), and the Merkle cells and Ruffini endings to steadily maintained deformation of the skin (DC to 200 Hz) [73].

3.2.3.3 Thermoreceptors Another group of sensory receptors detects the temperature of the skin. These receptors can respond to both constant and fluctuating skin temperatures. In responding to constant temperatures, the receptors discharge impulses continuously to indicate the temperature of the skin. They are very sensitive to changes in the skin temperature. There are two types of thermoreceptors: cold receptors and warm receptors. The cold receptors have a peak sensitivity of around 25–30°C and are excited by dynamic downshifts in temperature. The warm receptors have a peak sensitivity of around 39–40°C and are sensitive to increases in skin temperature [73].

3.2.3.4 Nociceptors Nociceptors are another group of sensory receptors, which respond to noxious stimuli such as heating the skin, strong pressure, or contact with sharp or damaging objects. The receptors have relatively high thresholds, to function as warning devices that enable the organism to take protective action. There are two major types of nociceptor. The first type are A fibers that have myelinated axons conducting between 10 and 40 m/sec and these are best adapted to respond to mechanical stimuli. The second type are small A or C fibers, either nonmyelinated or thinly myelinated, and these respond to a diversity of stimuli: high (>42°C) or low (<10°C) temperatures, pain-producing chemicals, and high intensity mechanical stimuli. These nociceptors have enhanced sensitivity in flamed tissues and may then be excited by normally innocuous stimuli [73].

Recent studies in conscious humans by direct recording from single nerve fibers in peripheral nerves have confirmed that isolated activation of an individual sensory receptor can result in distinct sensory perceptions. Meissner corpuscles cause touch sensations, Merkle receptors generate pressure, and nociceptors evoke pain sensations. The encoding of specific sensory information is begun by these sensory receptors in the skin. The central nervous system makes its further analysis through neural pathways by transferring the information to the brain.

3.2.4 Neural Pathways and Responses

The neural signals from the nerve endings are passed to the brain to formulate sensation. The pathways to the brain depend on two major principles: the types of nerve fibers and the place of termination of the pathway in the cortex.

Different types of nerve fibers carry different types of information to the brain. The nerve fibers can be classified in a number of ways: (i) by the types of stimuli that excite them, (ii) by the way they respond to stimuli (slow- or fast-adapting), and (iii) by their receptive field (large, ill-defined or small, well-defined). The receptive field refers to the region of the skin which, when stimulated, causes responses in a particular neural fiber. By these classification criteria, the nerve endings responding to mechanical deformation in glabrous skin have been classified into four types: (i) rapid adapting fibers with small and well-defined receptive fields, (ii) slow adapting fibers with similar receptive fields, (iii) slow adapting fibers with large, ill-defined receptive fields, and (iv) rapidly adapting fibers with similar receptive fields [72].

The second principal is that the location of a nerve ending determines where its information goes to the brain, regardless of the type of fiber it represents. There are 31 pairs of nerves, through which all the sensory information from skin is passed on to the spinal cord. Through the dorsal roots, the nerve endings enter into the back portion of the spinal cord. For the head region, there are 4 cranial nerves collecting cutaneous information. The information gets to the brain by two main pathways, each of which carries different types of information. In Fig. 3-2, various aspects of these pathways are illustrated. The features of the two pathways are summarized in Table 3-1.

Table 3-1
Description and Features of the Pathways

Medial pathways (i.e. the medial lemniscus)

• large, quick
• receive inputs from large, myelinated, fast-conducting Aβ fibers terminating in corpuscular endings
• have fibers responding to touch, temperature, and movement
• ascend the spinal cord on the same side of the body until they reach the brain stem
• most nerve fibers cross over to the other side at the brain stem
• continue to the thalamus
• finally arrive at the somatosensory cortex, which is located in the parietal region of the brain
• terminate in the samatosensory cortex on the opposite side of the body from where it started

Spinothalamic pathways

• slow, made up of many short fibers
• carry information about temperature and pain
• ascend on the opposite side of spinal cord from where input fibers terminate in the skin
• at the brain stem, divide into two branches:

Paleospinothalamic	Neospinothalamic
– specialized for signaling dull or burning pain	– specialized for signaling sharp or pricking pain
– receive inputs from small, unmyelinated, slow-conducting C fibers that terminate in free nerve endings in the skin	– receive most inputs from small, myelinated, slow conducting Aδ fibers
	– receive inputs from large, fast-conducting Aβ fibers

• innerveate several areas of the brain (e.g. thalamus and limbic system)
• then go to the somatosensory cortex

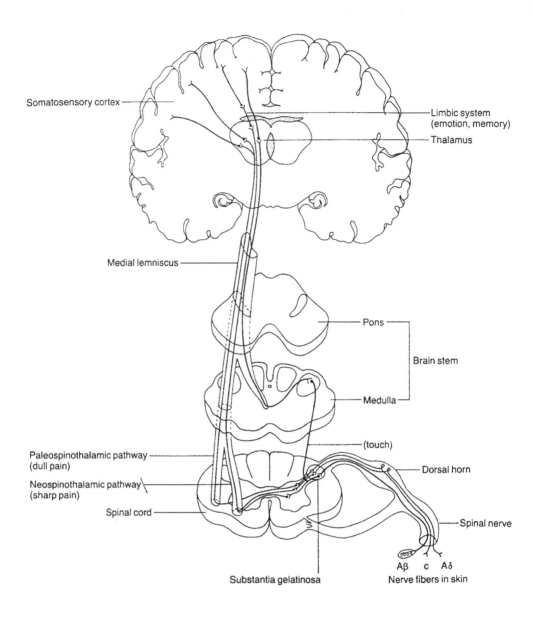

Somatosensory cortex

Limbic system (emotion, memory)

Thalamus

Medial lemniscus

Pons

Brain stem

Medulla

(touch)

Paleospinothalamic pathway (dull pain)

Dorsal horn

Neospinothalamic pathway (sharp pain)

Spinal cord

Spinal nerve

Aβ   c   Aδ
Nerve fibers in skin

Substantia gelatinosa

Fig. 3-2   Neural pathways from the skin to the brain [72]

There is a regular relationship between where a stimulus is applied to the skin and where neural activity occurs in the somatosensory cortex. Renfield and Rasmussen created a classic map of this relationship, shown in Fig. 3-3. This map was obtained by electrically stimulating the somatosensory cortex of patients during brain operations. As a point on the cortex was stimulated, the patients pointed out where they felt the sensation.

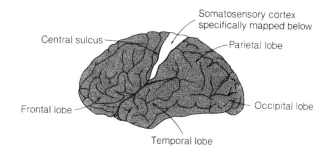

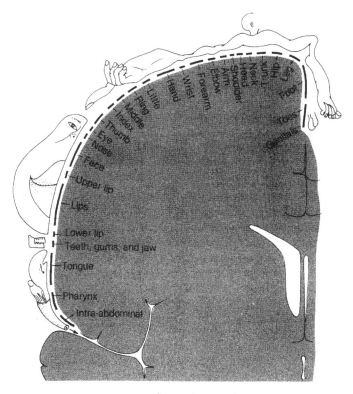

Sensory homunculus

Fig. 3-3   Renfield and Rasmussen's topographic map [72]

3.3 Perception of Sensations Related to Mechanical Stimuli

3.3.1 Dynamics of Wear Sensation

During wear, clothing contacts with the skin of most parts of our body, dynamically and continuously. This contact between skin and clothing has a number of features: (i) the areas of contact are large and cross over regions with various sensitivity; (ii) the body very often changes its physiological parameters such as skin temperature, sweating rate, and humidity at the skin surface, which generate various new thermal stimuli; and (iii) the body is often in movement, causing clothing to move towards and away from the skin frequently, which often induces new mechanical stimuli. These thermal and mechanical stimuli trigger responses from various sensory receptors and formulate various perceptions – touch, tactile, thermal, moisture, and more complex synthetic sensation, which affect the comfort status of the wearer.

3.3.2 Perception of Touch and Pressure

Any point on the surface of a human body can evoke the sensation of touch. However, the sensitivity varies from one region of the body to another. Fig. 3-4 shows the average absolute thresholds for different region of the female skin. The thresholds were obtained by applying a hair to the surface of the skin with different amount of force, and are expressed as the amount of force applied to the hair. The higher the bar, the greater the force needed to trigger the sensory receptor, and the lower the sensitivity. Obviously, the absolute thresholds vary considerably over the body surface. The threshold for touch sensation depends on both frequency of vibration of a stimulus and skin temperature. Each touch sensation seems to be located at a particular place on the skin and to be directly related to the amount of neural presentation at each area in the touch cortex [72].

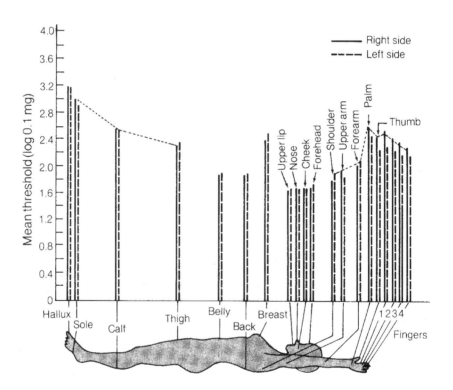

Fig. 3-4   Average absolute thresholds for different regions of the female skin

In the process of fabric–skin contact and mechanical interaction during wear, clothing will exert pressure and dynamic mechanical stimulation to the skin, which will in turn trigger various mechanoreceptors and generate a variety of touch sensations. Amano *et al.* (1996) [74] reported a study on the fluctuation of clothing pressure during wear by using spectrum analysis. They observed that the positions of a subject rather than the shape of the clothes affected the amplitude of the fluctuation. Under static conditions, clothing pressure fluctuated, with a peak at 0.2–0.4 Hz due to respiration. In the lower frequency range of the peak, the pressure spectrum changed with the clothes and the positions. They believed that the fluctuation in the clothing pressure was related to the comfort of clothing.

Sukigara and Ishibashi (1994) [75] measured the pressure between the fabric surface and the subject's hand during subjective assessments of fabric roughness. They found that an increase in pressure tends to produce stronger 'knobbiness' and 'roughness' evaluations. Momota *et al.* [76] studied the clothing pressure caused by Japanese women's high socks. They measured clothing pressure and sensory responses on subjects wearing such socks. After examining the relation between the clothing pressure and the sensory evaluation of the subjects, they concluded that, in order to design comfortable high socks that are not loose or wrinkled, it is necessary to ensure that: (i) the length of the socks is long enough to cover the upper calf; (ii) the stitches and patterns at the ankle can follow leg motion easily; and (iii) the pressure at the lower leg is within the range of 5–10 mm Hg at a standing position of rest.

Homota *et al.* [77] reported a study of the clothing pressure from wearing Japanese men's socks, by measuring the sock pressure on a foot model and on human subjects. They found that the sock pressure measured on human subjects was lower than that measured on the model foot. Subjects reported feeling comfortable at a pressure of around 10 mm Hg at the top and 5–10 mm Hg at the ankle.

In 1993, Makabe *et al.* [78] measured clothing pressures in the covered area of the waist for a corset and a waistband and recorded the sensory responses of subjects on the clothing pressure. They observed that the pressure at the waist was a function of the covering area, respiration, and the ability of samples to follow bodily movement. Subjects reported their perceptions on pressure at the waist line as: (i) no sense or no discomfort when the pressure was 0–15 gf/cm; (ii) negligible or slight discomfort when the pressure was 15–25 gf/cm; and (iii) extreme discomfort when the pressure exceeded 25 gf/cm.

Shimizu *et al.* (1993) [79] measured the clothing pressure on the body in a brassiere under static and dynamic conditions. They reported that the static pressure in the standing position was high at the shoulder, the side, and the back. Two main zones of pressure during movement were found at the shoulder and the back. Also, the pressure under static conditions was lower than that during movement. Makabe *et al.* (1991) [80] also studied clothing pressure from brassieres. They observed that wearer preferences were related to pressure distribution.

Shimizu (1990) [81] investigated the dynamic behavior of clothing on the knee of a person wearing slacks, using nine samples of slacks made from three kinds of fabrics with three different degrees of ease for three typical sequences of motion. They identified the high-pressure region by measurement of the dynamic pressure distribution over the knee surface, and found that the change in clothing pressure was a function of time.

3.3.3 Perception of Prickle, Itch, and Inflammation
Prickle is a sensation that is often complained of by consumers about next-to-skin garments, especially when fabric containing wool fibers is used for underwear garments. Prickle is usually described as the sensation of many gentle pinpricks. Traditionally, the prickle sensation associated with wool was considered associated with a skin allergic response. The degree of discomfort

caused by prickle varies with person to person and with the wear situations. Prolonged irritation that evokes scratching of the affected area can lead to skin inflammation. As these sensations have significant impact on the comfort experience of consumers in practical wear situations, substantial research has been carried out to study the mechanisms involved in fabric prickle sensations.

In 1984, Westerman *et al.* studied the relationship between sensations of prickle and itch, and human cutaneous small nerves [82]. Skin sensations were tested on the forearms of 12 volunteers, in whom anoxia nerve blocks of the forearm were produced by inflating a blood pressure cuff to 270 mm Hg on the upper forearm. As Fig. 3-5 shows, touch sensations were lost after about 20 minutes, but pain, temperature, and fabric-evoked prickle sensation remained until about 40 minutes. This result indicated that prickle sensations are associated with small nerve fibers.

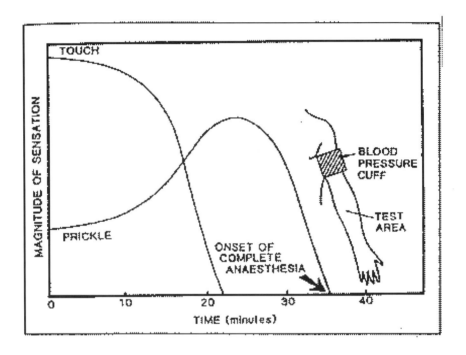

Fig. 3-5   Time course of loss of prickle, and touch sensations

Garnsworthy *et al.* [83,84] investigated the mechanisms of fabric prickle. They used invasive experiments on both human subjects and animals [85] to record the electrical activity of individual nerve fibers to determine which nerve fibers were responsible for the sensation of prickle. They reported that the neurophysiological basis for fabric-evoked prickle is not caused by skin allergic reaction, nor by chemicals released from wool. The cause of fabric prickle was identified as the mechanical stimulation of fabric to the skin that induces low-grade activity in a group of pain nerves, as shown in Fig. 3-6. As a fabric begins to contact the skin, the protruding fibers of the fabric will take all the force initially. As the body of the fabric moves closer to the skin, the forces increase and the protruding fibers bend. When the forces from the individual fibers reach certain levels, large shear forces in the skin are generated and pain nerve endings in the skin are activated.

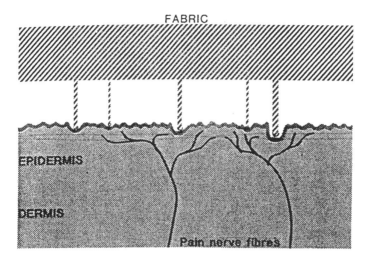

Fig. 3-6 Diagrammatic presentation of the mechanisms of fabric-evoked prickle sensation [86]

The nerve endings are identified as a group of pain receptors termed nociceptors (both Aä and polymodal C). The nociceptors that are activated cause a low rate of discharge from nociceptors over a wide area of skin. The critical buckling load that can trigger these pain receptors is around 0.75 mg or more at the point of contact with the skin. Fig. 3-7 shows the relationship between the responses from nociceptors and the density of high load bearing fiber ends on the fabric surface [83].

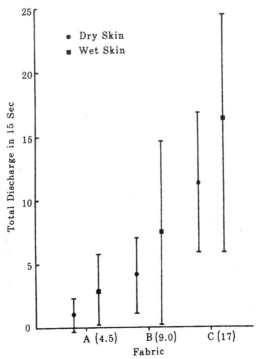

Fig. 3-7 Response of polymodal C nociceptors and fiber ends on fabric surface [83]

Garnsworthy *et al.* [84] investigated the psycho-physical relationship between the sensation of prickle magnitude and the measure of the physical stimulus of prickle by comparing the average forearm prickle test ratings with the density of high load bearing fiber ends on a fabric surface. The density was estimated by a method creating a permanent impression of the protruding fabric fibers in stretched polytetrafluouroethylene (PTFE) tape, which is described in detail in Reference [84]. An 11.4 cm² area of Teflon that had been impressed with the test fabrics was examined within 10 minutes of imprint. Three categories of fibers in the test fabric (<75 mg, 75–175 mg, and >175 mg) were counted by one of the authors who was unaware of which fabric he was evaluating. Thirteen fabrics were evaluated in such a way and were used for a subjective assessment of prickle sensation by 55 subjects. The mean subjective perception rating of prickliness was plotted against the prickle stimulus intensity (mean number of fiber ends exerting loads > 75 mg/10 cm²), as shown in Fig. 3-8. The data were analyzed using Steven's psychophysical power law. The fitted equation was:

$$R_p = 0.54 \; S_p^{0.66}$$

where R_p is the subjective sensation magnitude of prickle, S_p is the prickle stimulus intensity (number of protruding fabric hairs with buckling loads > 75 mg/10cm²). The correlation coefficient of the equation was 0.91.

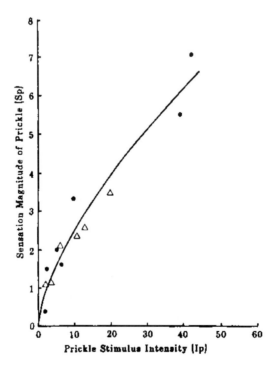

Fig. 3-8 Psychophysical relationship between prickle sensation and prickle stimulus intensity [84]

These author and others reported a number of findings that are important for understanding fabric prickle properties:

- It was found that summation of responses from the pain group of nerves is necessary for the initiation of pain sensation [87]. Prickle from fabrics could not be perceived if the density of high load bearing fiber ends is less than 3 per 10 cm^2 of the fabric, or the skin contact area is below 5 cm^2 [86].
- Sensitivity to fabric prickle is influenced by a number of factors:
    - (i) Males had higher thresholds and more variations in sensitivity to prickle than females;
    - (ii) Prickle sensitivity decreased progressively with age, since the skin is known to harden as age increases and prickle sensitivity decreases with hardness of the outermost skin layer.
    - (iii) Pain nerve endings are very close to the surface in hairy skin but not in glabrous skin, which explains why prickle cannot be felt with the fingers.
    - (iv) Prickle sensitivity increased with moisture content of skin, as water can soften the stratum corneum and allow the protruding fibers to penetrate more readily.
    - (v) Prickle sensitivity increased with ambient temperature in the range of 12–32°C at constant relative humidity (60–65%) as the skin moisture content increased due to perspiration in hot and humid conditions [86].

Itch is a sensation that has been shown to result from the activation of some superficial skin pain receptors [87]. The pain receptors responsible for itch may be of a different type to those responsible for the pricking sensation.

Skin inflammation (reddening) occurs in a small proportion of the population, as a consequence of prickle and itch resulting from mechanical stimulation of skin pain receptors from prickly fabrics — most likely through a mechanism termed axon reflex [88,89]. When some pain nerves in the skin are excited, vasoactive agents are released near the nerve endings. These substances dilate surface blood capillaries and cause reddening, first in the vicinity of activated pain nerve endings, and then spreading to larger areas of skin. The agents also promote pain nerve excitation and make the condition progressively worse, with itchiness and enhanced prickle sensitivity. A rapid flare response will usually be generated by this mechanism when a painful stimulus damages the skin. This is not an allergic response; rather it is an irritant response. Inflammation may occur rapidly (in minutes) or slowly (in hours). It can be relieved quickly after the fabric is removed from the skin, unless the fabric–skin contact is too long and has produced a severe reaction.

3.3.4 Roughness and Scratchiness

The sensations of roughness and scratchiness, which are related to the surface geometry, are perceived when a fabric moves across the skin. In the field of perception, extensive research has been carried out to study the physical and neural bases of roughness perception. To manipulate relevant physical features systematically and independently, two types of stimuli were created: (i) surfaces with regularly spaced ridges (gratings), and (ii) surfaces with raised dots. By using regularly spaced ridges, Lederman and Taylor [90–93] identified two major factors relevant to touch perception: (i) the spacing between neighboring ridges, and (ii) the amount of force applied. It was also observed that perceived roughness for a given surface was comparable, whether the finger moved over the surface or the surface moved over a stationary finger, when the applied forces were the same. This result suggested that the perception of roughness was only related to the cutaneous mechanoreceptors, and was independent of kinesthetic input.

Darian-Smith and Oke (1980) [94] investigated the responses of primate cutaneous mechanoreceptors to systematically applied grating stimuli, details of which are shown in Fig. 3-9. They found that the stimulus temporal frequency (cycles/sec), defined as the scanning velocity divided by the spatial period of the grating, was the primary factor related to the responses of all types of mechanoreceptors. They reported two important findings: (i) the spatial period of a stimulus cannot be related to the firing rate of individual mechanoreceptors, and (ii) each type of mechanoreceptor has a particular sensitive range of temporal frequency: Slow Adapting (20–60 Hz), Rapidly Adapting (60–200 Hz), and Pacinian Corpuscle (100–300 Hz).

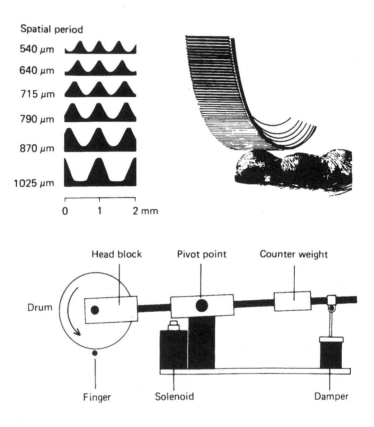

Fig. 3-9 Details of stimulator used for presenting gratings to finger pad skin [95, p.78, Fig. 16]

Goodwin *et al.* [96–98] reported a series of studies on the mechanoreceptors' responses to gratings stroked across their receptive fields. They reported that: (i) the response rate of all mechanoreceptors increased with increase in groove width from 0.18–2.0 mm, (ii) there was a small increase in ridge response rate with increasing ridge width, and (iii) the response rate decreased with increased velocity.

Greenspan and Bolanowski [99] summarized the psychophysical results from these findings, as shown in Table 3-2. As no responses from individual types of mechanoreceptors match the perceived roughness, it seems that the perception of roughness is not a function of any individual mechanoreceptor.

Table 3-2
A Visual Summary of Neurophysiological–Psychophysical Experiments on Roughness Perception [99]

Stimulus factor	Rapidly adapting		Pacinian Corpuscle		Slowly adapting		Psychophysical results
	ridge rate	cycle rate	ridge rate	cycle rate	ridge rate	cycle rate	Perceived roughness
Increasing groove width	/	/	/	/	/	/	/
Increasing ridge width	/	—	/	\	/	—	\
Increasing scan velocity	\	/	\	/	\	—	—

Connor *et al.* conducted a series of studies on the perception of roughness using dot spacing patterns. The major findings from these studies were:

- perceived roughness was correlated with the variability in mechanoreceptor response rates, particularly of slowly adapting afferent [100],
- roughness magnitude increased with larger dot separations [101],
- the between-fiber spatial variation in firing rates of slow adapting afferent provided a very good match with the shape of psychophysical function [101],
- perceived roughness was best related to differences in discharging rate over distances of 1– 2 mm on the skin [101],
- perceived roughness followed the trend of the slowly adapting mechanoreceptors, not the rapidly adapting ones [102].

These research outcomes provide the evidence that the perception of roughness is related to the spatial variations in responses between mechanoreceptors. The surfaces of textile products are very complex. Their surface spatial dot patterns and forces from protruding fiber ends vary considerably across different types of weave structure and fibers. LaMotte [103] studied the responses of mechanoreceptors to stroking the skin with nylon fabrics. He observed that response from rapidly adapting nerve fiber (Meissner) is related to the skin indentation, the weave pattern, the density of the fabric, and the rate of movement across the skin. Gwosdow *et al.* [104] reported that subjective perceptions of roughness increased with levels of skin wettness and were correlated with the frictional force required to pull fabrics across the forearms of subjects.

Behmann (1990) [105] made an attempt to relate subjectively perceived roughness to textile construction parameters. He reported that the irritation produced was provided by the roughness spacing. The roughness of woven and knitted fabric was found to be a function of the yarn diameter. Sukigara and Ishibashi [75] evaluated the surface roughness of polyester crepe fabrics in the grey and finished states, subjectively and objectively. Twenty female subjects were used. They found that the finished fabrics were perceived as being more 'knobby' and 'rough' than the grey fabrics. Roughness perception increased with the pressure between the fabric surface and the subject's hand. In 1995, Wilson and Laing [106] reported a study on the effect of wool fiber variables on tactile characteristics of homogeneous woven fabrics. They examined the extent of influence of wool fiber on the tactile characteristics of homogeneous fabrics with standardized fiber content, yarn structure, dye, and finishing treatment. Significant differences in the ranking of roughness and prickliness were found among these fabrics. The perceptions of roughness and prickliness were observed when the fiber diameters were > 34 microns.

Scratchiness is another sensory descriptor widely used by consumers to describe their experience of discomfort sensation with clothing. It has been found that scratchiness is highly related to the sensation of roughness in both consumer surveys [22] and the sensory responses of subjects in wear trials [41]. Mehrtens and McAlister [107] studied the subjective perception of scratchiness through wear trials under hot humid conditions by using garments made of fabrics with identical structural features but different fibers: Orlon, nylon, rayon staple fibers, and cotton. They found that scratchiness reception decreased as the filament flexural rigidity and friction decreased.

3.4 Perception of Sensations Related to Thermal and Moisture Stimuli
3.4.1 Functions of Thermal Sensitivity
Thermal sensations have long been recognized as an important aspect of comfort. Thermal senses tell us about our thermal state, both internal and external, and are indispensable to body temperature regulation and thereby to personal survival. Thermal sensitivity has three functions:

(i) body temperature regulation by either instrumental behavior (such as changing clothing) or autonomic behavior (such as vasoconstriction, vasodilation, sweating, and shivering),
(ii) avoidance of local damage to the skin from burning by cold or hot temperatures, and
(iii) monitoring temperature during touching an object [108].

The influence of thermal sensitivity on thermoregulation and overall thermal comfort will be discussed in the next section. It will focus on the transient cold and warm sensations as perceptions influencing the comfort status of the wearer.

3.4.2 Thermoreceptors
There are several types of thermoreceptors. Some thermoreceptors are located in the hypothalamus, spinal cord, and gut. Their main function is to monitor body temperature and involve the autonomic functions. Near the body surface, a variety of receptors respond to temperature, including 'warm' and 'cold' receptors, nociceptors, and SA (slowly adapting) mechanoreceptors [109].

Hensel [110] defined the general properties of specific cutaneous thermoreceptors as: (i) having a static discharge at constant temperature (T); (ii) dynamically responding to temperature changes (dT/dt) with either a positive temperature coefficient (warm receptors) or a negative temperature coefficient (cold receptors); (iii) not being excited by mechanical stimuli; and (iv) being active in the non-painful or innocuous temperature ranges.

Thermoreceptors can be classified into cold receptors and warm receptors, according to their dynamic responses. The static and dynamic response curves of cold and warm receptors are shown in Fig. 3-10. For static temperatures, cold receptors respond to temperature from –5°C to 43°C, with a peak at about 25–27°C [110]. Also, they respond to very high temperatures, above 45°C (which accounts for the paradoxical cold sensation that human beings experience when skin is exposed to very high temperature). Warm receptors start to discharge at constant temperatures beginning at 30°C. They increase their activity when the temperature rises and this peaks between 45°C and 47°C in the human hand, and between 41°C and 43°C for other populations of warm fibers in human skin [110]. For temperature changes, a warm receptor responds with an overshoot of its discharge on sudden warming and a transient inhabitation on cooling. On the other hand, a cold receptor responds with an inhabitation on warming and an overshooting on cooling.

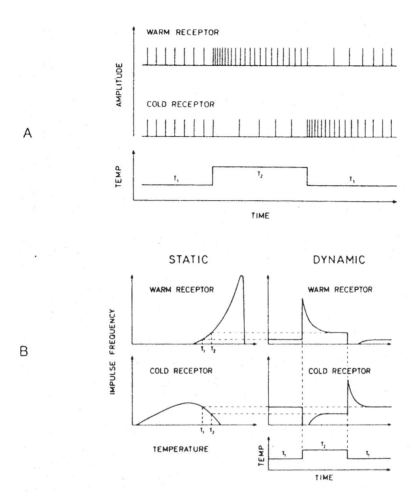

Fig. 3-10 The response curve of cold and warm receptors [110, p.35, Fig. 4.1]

In the skin of human subjects, cold receptors have been identified as being the radial nerves in hairy skin and warm receptors as being the radial nerves in hand dorsum through electrophysiological methods by Hensel *et al.*, [110]. In describing thermoreceptors, Spray [109] stated: 'The actual sensory receptors are apparently mitochondria-rich conical or bulbous projections of nerve fibers into the cytoplasm of basal epithelial cells. These projections are branched unmyelinated processes from afferent fibers that are commonly dichotomized into the A delta category of myelinated axons for cold receptors and C fibers for warm receptors, ... Specific thermoreceptor axons are smaller than those of temperature sensitive mechanoreceptors, which are recruited into the afferent responses at higher stimulus strength;...'

Regarding the neural pathway, Spray says: 'Primary thermal afferent from skin receptors have somata in representative sensory ganglia and project to the marginal zone of the spinal cord, synapsing on rostral brain stem and thalamic nuclei and projecting to somatosensory cortex. Integration at the first level of synaptic interaction is apparently considerable; ... The brain stem areas involve several midbrain raphe nuclei; thalamic areas of thermoreceptor representation include VB and NPT. Relay to the hypothalamus, where there is integration with information from deep body thermoreceptors, is apparently from midbrain raphe nuclei parallel to the thalamic projection' [109].

In addition to these thermoreceptors, nociceptors which are called 'heat receptors' respond to noxious temperature and are associated with pain. The SA mechanoreceptors also respond to temperature, which may reflect the psychological phenomena when we perceive simultaneous thermal and mechanical stimulation of the skin.

3.4.3 Fabric Thermal Sensations

Thermal sensations result from the responses of thermoreceptors to constant temperatures and transient transfer of heat from the skin (a cooling process) or to the skin (a warming process) after the fabric is placed on the skin. On the basis of Hensel's work, Ring and de Dear [111] described this neurophysiological response mathematically. The impulse frequency of a cold receptor or a warm receptor is a function of static temperature and the rate of temperature change:

$$Q(y,t) = K_s T_{sk}(y,t) + K_d \frac{\partial T_{sk}(y,t)}{\partial t} \qquad (3\text{-}1)$$

where $Q(y,t)$ is the pulse output response of a thermoreceptor as a function of time t and the depth y of a thermoreceptor in the skin, in impulses/s; K_s is the static differential sensitivity related to the steady state temperature; K_d is the dynamic differential sensitivity associated with temperature change; and $T_{sk}(y,t)$ is the temperature in the skin at thermoreceptor depth y as a function of time. The first term on the right of this equation represents the static discharge of the cutaneous thermoreceptors at constant temperature. The second term gives the dynamic response of thermoreceptors to temperature change. K_s and K_d are derived from the slope of the static and dynamic frequency curves ($\Delta F/\Delta T$), i.e. the change in frequency (ΔF) for a small change in temperature (ΔT). Hensel and Kenshalo [112] reported that a maximum static differential sensitivity (K_s) of -1 s^{-1}°C^{-1} and a maximum dynamic differential sensitivity (K_d) of -50°C^{-1} were found for the cold receptors in a cat's nose. Later, Kenshalo [113] reported that the average differential sensitivity of a cold fiber population in a monkey's skin was around -21°C^{-1}.

Konietzny and Hensel [110] found that the highest static differential sensitivity of human warm fibers was about 4°C^{-1}. The average maximum static frequency from a warm receptor in a cat's nose was 36 s^{-1}. The highest dynamic frequency was 200 s^{-1}. The dynamic differential sensitivity was found to be 70°C^{-1} [112].

These findings defined the existence of warm and cold receptors and provided the neurophysiological basis for the thermal perceptions. However, we still cannot state that the responses from these receptors determine the subjective perceptions of thermal sensations such as warmth and coolness. For static temperature sensations, Dykes (1975) [114] found that the average discharge frequency of cold fibers in a monkey's hand correlated well with static cold sensations in humans in the range between 34 and 25°C. For dynamic temperature sensations, it was found that the instantaneous frequency was not a satisfactory indicator and the number of impulses integrated over a certain time period must be considered [110]. Kenshalo (1976) proposed a 'central integrator' that has a long time constant of decay. The time constant was defined in such a way that the amount of decay is equal to the difference between the total impulses produced by fast and slow temperature changes [113].

Hensel described the concept of a 'central threshold' on the basis of the fact that individual cold and warm receptors may be excited without conscious thermal sensations. The threshold of cold sensation responds to an instantaneous frequency of 80 s^{-1} or to a total number of 120 impulses in a single cold fiber. The threshold of warm sensation seems to correspond to an average instantaneous frequency of 9 s^{-1} or to a total number of 28 impulses in a human warm fiber. The total number of impulses from a human single warm receptor at the threshold of conscious warm sensation is a function of the rate of temperature change, as shown in Fig. 3-11 [110].

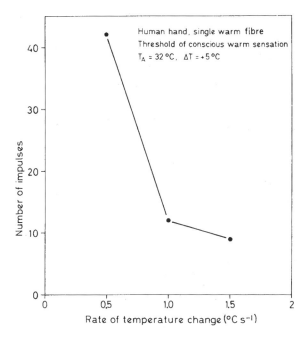

Fig. 3-11 Conscious warm sensation threshold and the discharge from a warm fiber in a human subject [110, p.86, Fig. 6.3]

These works suggested that the peak frequency and the cumulative impulses during the period of time following stimulus onset could be responsible for the way in which thermal responses are perceived. On the basis of these observations, Ring and de Dear [111] proposed that the intensity of thermal sensations (termed as psychosensory intensity, PSI) are proportional to the cumulative total impulses from stimulus onset at the thermoreceptor until such time as the receptor firing rate has decayed to within one impulse per second of the post stimulus steady state.

Li *et al.* [115] investigated the mechanisms of fabric coolness to the touch. The predicted temperature responses of skin to the contact of four fabrics were plotted (Fig. 3-12). Then, these temperature curves were used as boundary conditions at the skin surface for insertion in the thermoreceptor response model developed by Ring and de Dear [111]. The responses from the cold receptors as each of the four test fabrics was brought into contact with the skin were predicted from the model, as shown in Fig. 3-13. The integrals of the thermoreceptor frequency output curves (PSI values) were found to be 37.8 for wool, 29.4 for cotton, 20.3 for wool/polyester blend, and 8.3 for polyester.

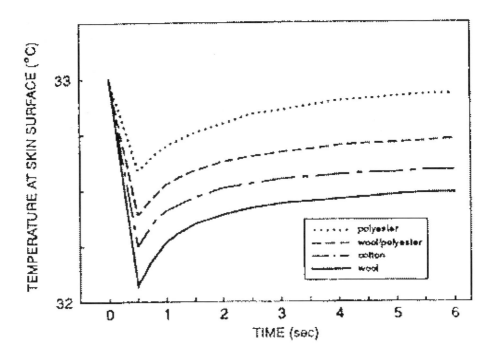

Fig. 3-12 Prediction of skin temperature change during fabric–skin contact [115]

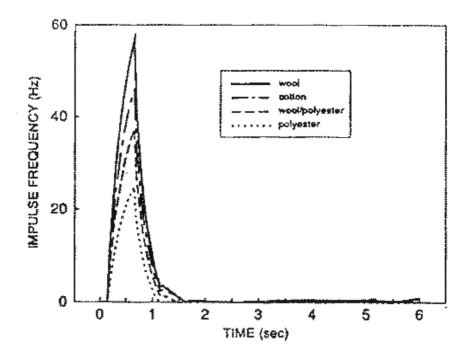

Fig. 3-13 Prediction of responses from cold receptors [115, p.591, Fig. 5]

The rank order of the impulse frequency of the cold receptors corresponded with the coolness ranking observed in the subjective trials conducted by Schneider *et al.* [40]. Results from the subjective perception trials were compared with the difference of the predicted PSI values between the pairs of fabrics in the last two columns of Table 3-3. Obviously, there was a good correspondence between the magnitude of the subjective differences and the predicted difference in coolness intensity, indicating that the integral value of the output frequency from skin cold receptors is a good predictor of fabric coolness.

Table 3-3
Comparison of PSI and Subjective Coolness Ranking

Comparison A	B	A cooler than B	Predicted PSI difference
Wool	polyester	15	29.5
Wool	wool/polyester	14	17.5
Wool	cotton	6	8.4
Cotton	polyester	14	21.1
Wool/polyester	polyester	15	12.0

3.4.4 Dampness Sensations
Moisture in clothing has been widely acknowledged as one of the most important factors contributing to discomfort during wear. Nielsen and Endrusick (1990) [116] observed that the sensation of humidity is correlated with skin wetness. They recommended the use of the subjective sensations of wetness of skin and clothing as a sensitive tool to evaluate the thermal function of garments.

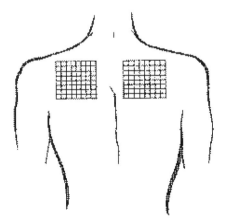

Fig. 3-14 Mapping of the right and left scapular regions for sensitivity to moisture

Sweeney and Branson (1990) [117] used a psychophysical approach to study the assessment of moisture sensation in clothing. They applied a method of constant stimuli to obtain the absolute and difference thresholds of moisture sensation at the upper-back area of volunteers (see Fig. 3-14). Subjects were trained to respond to the sensation of moisture, not temperature. By adding water to fabric swatches of one type of fabric as stimuli, they found that the absolute threshold of moisture sensation was 0.024 ml and difference threshold was 0.039 ml. Also, the relationship between the proportion of moisture detection and stimulus intensity exhibited a linear function.

In a further work, Sweeney and Branson (1990) [44] employed a magnitude estimation method to study the relationship between moisture stimulus intensity and moisture sensation. Wetted fabric swatches were applied to the upper back to obtain moisture sensation as before, using one type of fabric with seven moisture levels. During the experiment, Stevens' method of magnitude estimation was used to instruct subjects to estimate their perception of moisture. Thirteen subjects were tested. Stevens' power law was applied to obtain the relationship between the moisture magnitude ψ and the moisture stimulus ϕ, as shown in Fig. 3-15, where $\psi = 31.62\ \phi^{0.53}$, and the square of correlation coefficient (r^2) is 0.96.

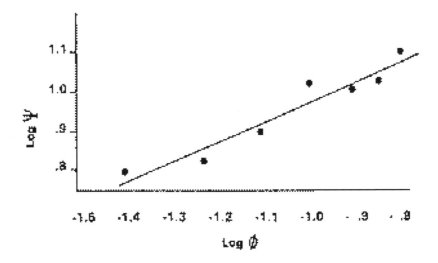

Fig. 3-15 Psychophysical relationships between moisture perception (ψ) and moisture content (ϕ) [44]

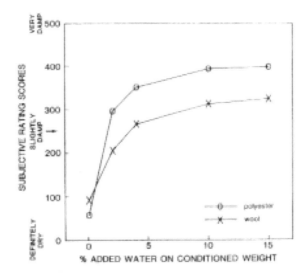

Fig. 3-16 Subjective perception of fabric dampness [33]

Li *et al.* (1993) [33] and Plante *et al.* (1995) [34] reported a series of studies on the perception of fabric dampness, using a sliding interval scale ranging from 'definitely dry' to 'very damp'. Fabrics with different levels of hygroscopicity were studied at five levels of moisture content and different levels of ambient relative humidity. The relationships between dampness perception and fabric moisture content also exhibited a power function, as shown in Fig. 3-16. In a further study, it was found that the perception of dampness depended not only on fabric moisture content, but also on fiber hygroscopicity and ambient relative humidity. With the same water content above equilibrium regain, the perception of fabric dampness decreased with fiber hygroscopicity at low relative humidity and seemed to decrease with relative humidity, particularly with lower hygroscopic fibers [34]. This casts a doubt on whether fabric moisture content is the actual stimuli for moisture sensation.

Furthermore, the neurophysiological basis of moisture perception is still not clear up to now. Clark and Edholm [118] stated in a monograph that the general consensus of opinion is that there are no specific moisture detectors in human beings. Kenins *et al.* [119] conducted a series of experiments to study the mechanisms of human perception of air humidity by using the human forearm and hand. Their conclusions were that the experimental results provided little evidence to support the assumption of a specific humidity detector in humans and that humidity might be perceived through some indirect mechanisms.

Bentley (1900) [120] reported that a close-fitting garment applying even pressure with a cold temperature could produce a feeling of wetness in the absence of moisture. This finding suggested that the dampness sensation might be a synthetic sensation that consists of a number of components such as fabric temperature, pressure, and distribution of pressure during the contact between skin and fabric. This hypothesis is supported by the observation that dry fabrics (equilibrated to test conditions) were never rated as 'definitely dry' by subjects. The perceived dampness with these dry fabrics increased with fiber hygroscopicity, which is consistent with the observation on fabric coolness perception.

These observations excited an investigation on the physical mechanisms of fabric dampness perception. Li *et al.* (1993,1995) [33,121] conducted a series of physiological experiments and mathematically simulated the heat and moisture processes in the fabrics and between the fabric and skin during the fabric–skin contact. They identified that the skin temperature drop during the contact plays a key role in the perception of dampness. The relationship between subjective dampness rating and skin temperature drop exhibited a power function, as shown in Fig. 3-17.

3.5 Perception of Texture

According to Fiore and Kimle [122], texture is the uniformity and variation of the surface of an object, which can be the description of actual or implied features of surfaces. The texture of a surface can be described in many ways, such as *smooth*, *rough*, *shiny*, or *dull*. Obviously, texture is a complex synthetic sensation that covers many aspects of the sensory features of a surface, including visual, auditory, and various tactile perceptions. In normal situations, people often see a garment or fabric when they touch or wear it, and hear the sound of the contact and friction between the skin and the fabric or between different parts of the clothing material. All the information gained from the visual, tactile, and auditory signals provides us with the overall perception of a surface texture.

Perceptions of tactile texture are derived from the activation of sensory receptors in the skin, not only in the hands but all over the body surface, during the contact between the skin and clothing. Texture sensation may involve all three types of sensory receptors: pain, touch, and temperature.

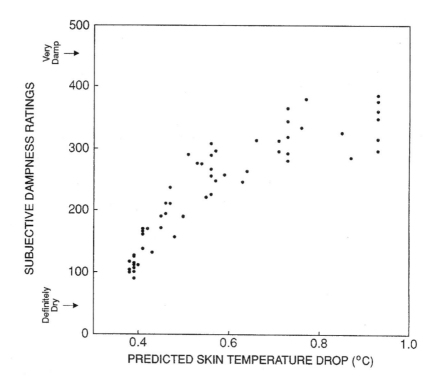

Fig. 3-17 Psychophysical relationship between dampness perception and predicted skin temperature change [121]

Touch is most likely the main component of texture, and roughness is an important aspect. Therefore, a large number of research papers have been published on the neurophysiological basis of roughness perception, as discussed in the previous section. Lederman (1982) reviewed the variables involved in tactual texture perception. He noted that the judgment of texture is dependent on the spatial distribution of the texture elements from moment to moment and on the rate of the fingers' travel over the texture elements [123].

Visual perception is a very important sensory channel in assessment of texture of a fabric or garment. In certain dimensions, vision plays a dominant role in the perception of texture. For instance, vision seems dominant over touch for identification of two-dimensional forms, spatial features, and color [71]. On the other hand, we learn how to recognize different texture and make assumptions about how they feel through past experience of how different surfaces and textures looked and felt.

Sometimes, intersensory conflict may occur when the visual perception is distorted, or the stored impression does not apply in a particular situation. For instance, visual perception of a snake's skin may imply that it is slick and slimy, but, when touched, the surface is dry and smooth. Normally, the senses operate in a cooperative manner. There may be a division of labor between vision and touch. Vision often provides guidance for the hand, with which touch is used to gain information about surface characteristics such as texture, hardness or softness, and thermal properties [71].

Brown *et al.* [124] reported a study on the effect of sensory interactions on descriptions of fabrics. In examining how sensory interactions may affect perceptual responses to fabrics, 169 subjects were asked to sort 24 fabrics into groups, based upon how they felt the fabrics. Subjects were divided into two groups: one group touched and viewed the fabrics (sensory interaction group), the second group touched the fabrics only (sensory isolation group). Content analysis of the subjects' written explanations of their sorting methods categorized their descriptive terminology according to texture, fabric traits, fabric name, fiber content, fabric weight, end use, appearance, extended inferences, and affective responses. The authors found significant differences between the terminology of the sensory interaction and sensory isolation groups. Whereas subjects in the sensory interaction group were more likely to use terms classified as end-use and appearance terms, subjects from the sensory isolation group were more likely to use terms classified as texture and fiber content terms.

Dallas *et al.* (1994) [125] reported their investigation on sensory perceptions of fabrics in relation to gender differences. Male and female subjects sorted 60 fabric samples into groups based on tactile only or visual and tactile sensory interaction. The descriptive terms and phrases used by the test subjects when explaining why they grouped particular fabrics together were recorded and analyzed. They observed that the adjectives used were primarily related to fabric hand and texture or fiber/fabric structure, and that, although males and females possess a common set of descriptive terms for textiles, they differ in the frequency and variety of their sensory interaction and perceptual responses to fabrics.

In 1995, Jung and Naruse [126] used the ranking method and the paired-comparison method to evaluate the surface qualities of silk fabrics. Eight types of silk fabric with the same material density and yarn counts, but different weaves, were used. The fabrics were evaluated visually in respect to luster, quality of texture, and attractiveness, with white and black samples. It was observed that luster and attractiveness perception was considerably affected by color, but surface roughness and thickness were not.

3.6 Perception of Fabric Hand

Fabric hand (or handle), which describes the way a fabric feels during the touch by a human hand, is an important aspect of fabric texture [127]. Fabric hand properties have been extensively studied in the areas of subjective sensory descriptors, psychophysics, fabric mechanics, and objective and subjective assessment methods. Fabric hand is a complex synthetic sensation that consists of many dimensions, which are obtained through the active manipulation of a fabric by a human hand. To understand the mechanism of fabric hand, it is necessary to explore the neural sensory characteristics of our hands.

Our hands are the essential tools for us to explore and manipulate the external world, for which their sensory function plays a very significant role. The activities of exploration and manipulation are based on the effective integration of movement and sensory perceptions. The hand has several essential sensory mechanisms, including muscle sense and kinesthesia. It has been found that the mechanoreceptors in the glabrous skin play a key role. Human hands are equipped with large number of nerve endings, having about 17 000 units that are sensitive to non-noxious mechanical deformation of the skin in the glabrous skin of one hand. A sensory unit is made up of a nerve cell

Table 3-4
Sensory Receptors in a Human Hand [128]

Types of receptors	Type I		Type II	
	Fast adapting I	Slow adapting I	Fast adapting II	Slow adapting II
Type of end organ connected	Meissner corpuscles	Merkle cells	Pacinian corpuscles and Golgi-Mazzoni bodies	Ruffini endings
No. of end organ connected	multiple	multiple	single	single
Responding stimuli	moving stimuli	dynamic sensitivity to moving stimuli, stationary stimuli	skin deformation, rapid moving stimuli, remote stimuli	direct skin deformation, stretching of skin, tangential forces in the skin
Location	basal layer of epidermis	basal layer of epidermis	in subcutaneous tissue	below basal layer of epidermis
Distribution	140 cm^2–fingertip 5–10 time lower in palm	70 cm^2–fingertip 5–10 time lower in palm	uniform in whole glabrous skin 25 cm^2	uniform in whole glabrous skin 25 cm^2
Proportion	~70% of the sensory units		~30% of the sensory units	
Functions	spatial discrimination	skin indentation		skin indentation

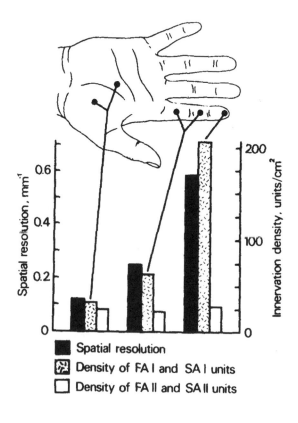

Fig. 3-18 Spatial resolution capacity and density of sensory units in the hand [128]

with a long axon running in the peripheral nerve and the associated organ(s) in the skin or subcutaneous tissues. These sensory units constitute the sensory basis for the ability of spatial and temporal discrimination in the skin area of the hand [128]. In these units, there are four types of sensory receptors with distinctly different response characteristics, which are summarized in Table 3-4.

The Type I receptors have very high densities at the fingertips, as shown in Fig. 3-18, and a number of features, including the following:

- They have small receptive fields with distinct borders to form a fine-grain system, allowing accurate location of stimuli;
- The two-point discrimination at the fingertip is close to the theoretical limit;
- The distinct receptive field of the sensory units has a counterpart in the receptive space of the mind, i.e. an individual afferent unit may be accurately represented within the brain;
- For the fingertips that are represented at the macula of the somatosensory system, the threshold is set by the sensitivity of the skin receptors, implying that the brain extracts information out of a neural quantum in the peripheral nerve;
- In other regions of the hand, the threshold is set by mechanisms within the brain [128];
- A single impulse from a Meissner unit may produce a sensation of touch, whereas 10–20 impulses from a Merkle unit may be required to produce such a sensation.

There is a fundamental difference between the perception of touch by wearing a garment and by handling a fabric. When wearing a garment, touch is passive, where the wearer does not move intentionally to get information about the clothing; information is essentially imposed on the skin. In the process of handling a fabric, touch is active, where the observer is moving hands intentionally to obtain objective information about the fabric. According to Gibson [129], fabric handle is active touch. Heller [71] also discussed the differences between active and passive touch and he distinguished *synthetic* touch with *analytic* touch. Synthetic touch is used to obtain an overall impression by a resting hand, while analytic touch is aimed to gain exhaustive information about an object's features.

Katz [130] classified active touch into four categories: (i) gliding touch – short back-and-forth motion of the hand to gain information about surface variations or texture; (ii) sweeping touch – using one or more fingers scanning across an object's surface to obtain information regarding contours, edges and geometrical relationships of parts; (iii) grasping – like gliding or sweeping, and using the thumb to get information about two or more surfaces simultaneously; finally (4) kinematic grasping – comprehensive exploration of an object's features.

Gibson (1962) [129] pointed out that, even though during active touch exploratory movements seem very variable, they are not aimless. The aim of the ever-changing movement is to isolate and enhance the component of stimulation that specifies the features of the object, and the nature of the component is to detect and isolate invariance in changing stimulation. Following this argument, Davidson *et al.* (1972) outlined the objectives of active touch activity as: (i) to link accurate and inaccurate methods of scanning to specific attributes under investigation, and (ii) to derive the characteristics of effective scanning methods, which are determined by how well the scanning strategy focuses on the relevant stimulus attribute and how well it can be encoded [131,132].

In an attempt to identify the features of active touch activity, Davidson *et al.* studied the scanning strategy by asking subjects to make matching judgments of replicas of three-dimensional free-form shapes. By videotaping the scanning activities, they found four types of exploratory movements: (i) global search – independent use of fingers to examine several stimuli simultaneously, (ii) detailed search – coordinated use of fingers to focus on single aspects of the stimulus, (iii) palms search – pressing the palm down onto the top of the stimulus, and (iv) tracing – moving fingertips along stimulus contours. They found that subjects used detailed and global strategies the most, but used other strategies more as the inspection time increased [131].

Lederman and Klatzky [133] conducted a series of experiments to examine 'exploratory procedures', which were defined as stereotyped patterns having certain invariant characteristics used to obtain information about a specific object property. By assessing the performance of subjects who used an exploratory procedure spontaneously or used one under instruction, they found that the procedures chosen by subjects produced the optimal performance to obtain the information desired. Four types of exploratory procedures were identified: (i) 'pressure' – using the hand to exert force against the object for hardness, (ii) 'lateral motion' – sideways movement between skin and object for texture, (iii) 'enclosure' – simultaneously contacting parts of the hand to mold the hand to the object for global shape, and (iv) 'contour following' – smooth and repeated hand movements along a contour segment for exact shape.

On evaluation of fabric handle, Bishop [54] pointed out that human subjects usually know instinctively how to manipulate the fabric to obtain the required information when they are asked to assess it against particular attribute descriptors familiar to them. However, their instincts vary considerably among subjects. They usually use different methods to assess the given attributes and may assess different features against the same descriptor. Therefore, he thinks that it is necessary to discuss both the meaning of the descriptor and the handling method with subjects.

Elder *et al.* [42] reported a detailed description of procedures for evaluating fabric softness, in which the method of how to apply pressure on a fabric sample and the amount of force to be applied were specified clearly. This methodology was adopted by Mackay [43] to train consumers for evaluating ten different fabric attributes.

Bishop [54] reviewed the fabric-handling techniques in the literature and agreed with the statement by Elder *et al.* [42]: 'little detail appears in the literature as to the instructions given to judges with regard to the method to be used to handle samples'. Generally, it has to be assumed that in most cases judges are allowed to handle fabric samples 'as they see fit'. This approach adopted by textile researchers may have good underlying reasons, according to Gibson's and Davidson's argument, and the observations from Lederman's experiments, provided that the invariants (features) under consideration are clearly explained to the subjects. However, it seems that fundamental information on the relationships between the sensory unit distribution, hand movement, and the features of the fabric under assessment is still missing.

The psychophysical relationships of subjective perception of softness and stiffness with fabric compression, percentage compressibility, flexural rigidity, coercive couple, and drape coefficient were studied by Elder *et al.*, using Steven's power law [42,60,134]. They found that the significance of Steven's power law relationship varies with different combinations of perception and objective properties and with different types of fabrics. The relationship is significant between softness and compression for both woven and non-woven fabrics. It is significant between subjective softness and percentage compressibility for woven fabrics, but not for non-woven fabrics. The relationship is significant between stiffness and coercive couple for the non-woven fabrics but not for woven fabrics, and is significant between stiffness and drape coefficient for woven fabrics. These observations indicated that the psychophysical relationships between subjective handle perception and fabric objective properties are complex in nature and are related to the features of different types of fabrics.

Hu and co-workers [135] studied the psychophysical relationships between the subjective perception of primary hand (stiffness, smoothness, fullness, and softness) and various fabric mechanical properties (bending, shearing, tensile, compression, surface, thickness, and weight) measured on the Kawabata Evaluation System (KES). The subjective primary hand values were obtained by asking subjects to rank 39 fabrics against the linear scale of the Japanese HESC standards. Hu *et al.* related individual subjective handle perceptions to the sum of contributions from a number of fabric mechanical properties by using a multiple regression method. Four models were compared, including linear function, the mixed linear and log-linear function used by Kawabata, Webber-Fechner law, and Steven's law. They observed that Stevens' law had the smallest deviations among the four models. Bishop [54] pointed out that, in comparison with the correlation coefficients of the psychophysical equations from Elder's work, the equations obtained seem disappointing. He suggested that the use of psychophysical laws (Stevens' or Weber-Fechner law) is not appropriate unless the subjective hand values are represented by the judges' estimates of their own responses to the fabric stimuli.

These results provide a good estimation of the possible psychophysical relationships between subjective hand perception and fabric mechanical properties. However, they do not give us a sound understanding of the neurophysiological and psychophysical mechanisms involved in fabric handle perception. Further research needs to be carried out to answer a number of fundamental questions, such as:

- How do the active touch movements of the hand generate various types of mechanical stimuli to the touch receptors?
- How do the four types of mechanoreceptors respond to the mechanical stimuli?
- How are the neurophysiological responses from the receptors coded and transferred to the brain?
- How does the brain process the information and formulate various subjective hand perceptions?

By answering these questions, a sound understanding of the hand perception might be established.

4. THERMAL PHYSIOLOGY AND COMFORT

4.1 Clothing and Thermal Comfort

One of the fundamental functions of clothing is to keep the human body in an appropriate thermal environment in which it can maintain its thermal balance and thermal comfort. During the course of biological development, the human body has lost much of its ability to control heat loss and maintain thermal balance. Therefore, clothing is needed to protect the body against climatic influence and to assist its own thermal control functions under various combinations of environmental conditions and physical activities. In such a way, the body's thermal balance is achieved and a microclimate which is perceived to be comfortable, is created next to the skin. In other words, an important task of clothing is to support the body's thermoregulatory system to keep its temperature within a median range, even if the external environment and physical activities change in a much broader range.

An understanding of the role of clothing in the thermal balance of the human body and thermal comfort under steady-state conditions has been developed over the past few decades, and this has been widely used in the clothing industry and the heating–ventilating industry. The human body is rarely in a thermal steady state, but is continually exposed to transients in physical activity and environmental conditions. Hygroscopic fibers such as wool and cotton absorb moisture vapor from ambient air when humidity rises, and release heat. Similarly, when the humidity falls, moisture is released, and heat is taken up by the fibers. Under transient conditions, this sorption behavior of fibers can play an important role in the heat exchange between the human body and the environment, and in thermal comfort perceptions.

4.2 Thermal Comfort

The thermal comfort of man depends on combinations of clothing, climate, and physical activity. Yaglou and Miller (1925) [136] defined 'effective temperature' as an index of warmth perception when a human body is exposed to various temperatures, humidity, and air movements. The scale of effective temperature was determined by the temperature of still, saturated air which was felt as warm as the given conditions. For instance, any ambient condition has an effective temperature of 60°F when it is perceived as warm as still air saturated with water vapor at 60°F. Effective temperature was adopted by the American Society of Heating and Ventilating Engineers as the operating scale for establishing comfort charts for clothed sedentary individuals exposed to a variety of temperatures, relative humidities, and wind velocities.

Rohles (1967) [137] derived an equation using multiple regression to predict thermal sensations after an exposure of three hours:

$$Y = 0.1509 T_{ab} + 0.01 H_a - 8.3719 \qquad (4\text{-}1)$$

where Y is the thermal sensation on the scale of 1 = cold, 2 = cool, 3 = slightly cool, 4 = comfortable, 5 = slightly warm, 6 = warm, and 7 = hot; T_{ab} is the dry bulb temperature in °F; and H_a is the relative humidity in percent.

Gagge *et al.* (1967) [138] studied the sensory comfort and thermal sensations of resting–sitting, unclothed subjects, and compared them with the associated physiological responses under steady-state and transient conditions of 12–48°C. When exposed to steady cold and warm environments, thermal comfort and neutral temperature sensations lay between 28–30°C when no physiological temperature regulatory effort was needed. Discomfort perception was related to lowering average skin temperature toward cold environments and increased sweating towards hot environments.

The same conclusion was drawn for transient changes — when the subjects were exposed from comfortable to uncomfortable, neutral to cold, and neutral to warm. Thermal discomfort was found to be an excellent stimulus for behavioral activity by man. Thermal sensation gave man an early anticipatory drive for conscious action to change his body's microclimate rather than depend on natural, but short-term, thermal protection by sweating, vasodilatation, or vasoconstriction and shivering.

In 1969, Gagge *et al.* [139] reported a study on comfort, thermal sensations, and associated physiological responses during exercise at various ambient temperatures. The authors concluded that, after 30–40 minutes of steady exercise, temperature sensations from *cool* to *hot* were mainly correlated with skin and ambient temperatures; warm discomfort was related to skin sweating and skin conductance. During steady state exercise, perception of temperature was dominated by sensory mechanisms in the skin, while warm discomfort was mainly determined by thermoregulatory mechanisms. The comfort and thermal sensations during thermal transients caused by the rise in metabolic rate at the start of exercise were correlated with the initial rise in mean body temperature.

Hensel (1981) [110], and Carterete and Friedman (1973,1974) [140,141] pointed out the physiological basis of thermal comfort, and the difference between thermal comfort and temperature sensations. Temperature sensations are mainly derived from cutaneous thermoreceptors, which are used to judge the thermal state of objects or the environment. Thermal comfort and discomfort reflect a general state of the thermoregulatory system, which is the integration of afferent signals from both cutaneous and internal thermoreceptors. Therefore, the measurements of temperature sensations and of thermal comfort need to be distinguished. McNall *et al.* [142] used two separate scales to study thermal sensations and thermal comfort, as summarized in Table 4-1.

Table 4-1
Scales for Thermal Comfort and Thermal Sensations

Thermal sensations	Thermal comfort
1. very cold	1. uncomfortably cold
2. cold	2. colder than comfortable
3. cool	3. much cooler than comfortable
4. slightly cool	4. slightly cooler than comfortable
5. neutral	5. comfortable
6. slightly warm	6. slightly warmer than comfortable
7. warm	7. much warmer than comfortable
8. hot	8. hotter than comfortable
9. very hot	9. uncomfortably hot

Fanger (1970) [4] developed a mathematical model to define the neutral thermal comfort zone of men in different combinations of clothing and activity levels. Mean skin temperature and sweat secretion rates were used as physical measures of comfort. Based on Fanger's work, the American Society of Heating, Refrigerating and Air Conditioning Engineers developed generalized comfort charts and indices of thermal sensation for predicting comfort acceptance under different combinations of clothing insulation, metabolic level, air temperature, and wet-bulb temperature (or radiant temperature) [6].

Fanger (1985) presented an international standard dealing with thermal comfort, ISO 7730, and discussed the philosophy and the scientific basis behind this standard [143]. It aimed to specify conditions that would be acceptable in thermal comfort to a given percentage of the population. In the standard, thermal comfort was defined as the condition of mind that expresses satisfaction with the thermal environment. Dissatisfaction, which may be caused by warm or cool discomfort for the body in general, is expressed by the PMV and PPD indices. PMV index is the Predicted Mean Vote, which is used to predict the thermal sensation for the body as a whole on a seven-point scale, from cold to hot. PPD index is the Predicted Percentage of Dissatisfaction. The ISO standard recommended a PMV within the range of –0.5 to +0.5, meaning that the PPD should be lower than 10%. When PMV is zero, the optimal operative temperature is achieved, this being a function of activity and clothing.

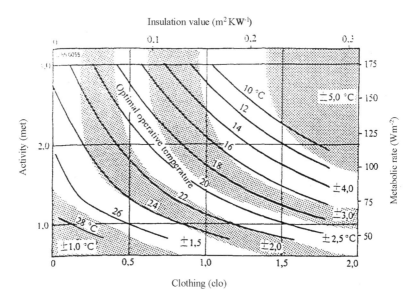

Fig. 4-1   Optimal operative temperature

Figure 4-1 shows the operative temperature, defined as the uniform temperature of an enclosure in which an occupant would exchange the same amount of heat by radiation and convection as in the actual non-uniform environment. For normal practical applications, the operative temperature is roughly equal to the mean value of mean radiant and air temperatures.

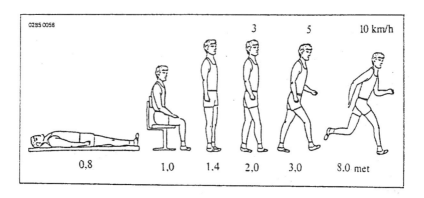

Fig. 4-2   Metabolic rates of physical activities

In Fig. 4-1, the curves show the optimal operative temperature which satisfies most people in given clothing and in a given activity. The shaded areas show the acceptable ranges around the optimal temperature. The metabolic rate can be estimated from physical activities, as shown in Fig. 4-2, while the thermal insulation of clothing (*clo*) can also be estimated from the type of clothing showing in Fig. 4-3. Using the metabolic rate and clothing *clo* values, the optimal effective temperature and its tolerance limit can be estimated from Fig. 4-1.

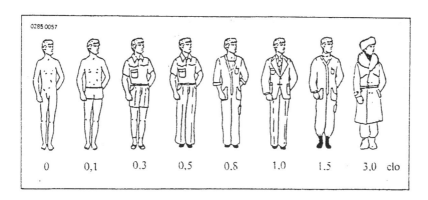

Fig. 4-3   Thermal insulation of typical clothing

Gagge *et al.* (1971) [5], as a rational starting basis, developed an environmental temperature scale based on the heat exchange equations during the passive state and the effect of physiological regulatory controls. The temperature scale used the Humid Operative Temperature (T_{oh}), defined as the temperature of an imaginary environment to which the body would lose the same heat by radiation, convection and evaporation as it would in the actual environment. A new 'effective temperature' scale was also constructed for a sedentary, normally clothed (0.6 *clo*) subject, on the basis of loci of constant wettedness caused by regulatory sweating. Further, in 1973, Gagge [144] defined three rational temperature indices: Standard Operative (T_{so}), Standard Humid Operative (T_{soh}), and Standard Effective (*SET*) Temperatures, in terms of average skin temperature, skin wettedness, and the associated heat transfer coefficients. Generally speaking, T_{so} is an index of thermal stress caused by the environment; T_{soh} is an index of thermal strain caused by T_{so}; and SET is an index temperature, describing the dry bulb temperature of the standard environment at 50% r.h. that causes the same heat exchange for the same T_{so}, skin wettedness, and average skin temperature.

In 1986, Gagge *et al.* [6] proposed another new Predicted Mean Vote index, designated PMV*, by simply replacing the operative temperature, T_o, in Fanger's Comfort Equation with *SET*. Gagge pointed out that Fanger's PMV is primarily based on heat load; it is not sensitive to changes in relative humidity or vapor pressure, nor to the vapor permeability of clothing worn. By defining PMV with SET instead of T_o, the new PMV (PMV*) is able to respond to the thermal stress by heat load, the heat strain by changing humidity of the environment, and the vapor permeability of the clothing worn.

4.3 Thermoregulatary Mechanisms of the Human Body

The human body has the ability to regulate its internal temperature with a certain level of accuracy, under the changes of external and internal conditions. The temperature regulation works through biological mechanisms — the specific central and peripheral nervous systems continuously detect the temperature fluctuations in the body and attempt to keep them in balance by means of biological actions.

Hensel[110] described physiological temperature regulation as a complex system that contains multiple sensors, multiple feedback loops, and multiple outputs. Fig. 4-4 shows his model of autonomic temperature regulation in man. The control variable is an integrated value of multiple temperatures, such as the central nervous temperature (T_{cn}), the extra-central deep body temperature (T_{db}) and the skin temperature (T_{sk}). Hensel defined the 'Weighted Mean Body Temperature' (T_{mb}) as the controlled variable for practical purposes:

$$T_{mb} = a\,T_i + (1-a)\,T_{sk}\ (a < 1) \tag{4-2}$$

where T_i is the internal body temperature and T_{sk} is the average skin temperature; a is the weighting ratio presenting the relative effect of T_{sk} and T_i in a linear control function. A value between 0.87 and 0.9 was proposed by measuring T_i in the oesophagus.

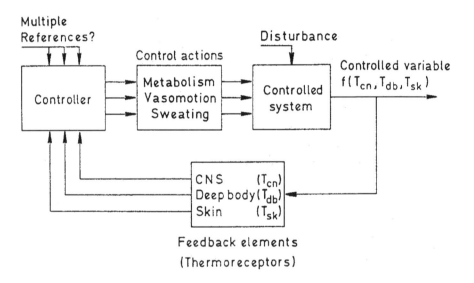

Fig. 4-4   Schematic diagram of autonomic temperature regulation in man [110, p.11, Fig. 2.2]

The references (or set temperatures) for different control actions such as metabolism, vasomotion and sweating, might be different. The heat-dissipation mechanisms, such as sweating driven by warm receptors, may have a higher set temperature than heat-production mechanisms driven by cold receptors. Therefore, there is a zone of thermal neutrality in which no thermal regulation occurs.

Hensel classified the thermal regulation mechanisms into three categories: autonomic regulation, behavior regulation, and technical regulation.

The autonomic regulation responds to thermal disturbances from internal heat generated by exercise and environmental heat or cold. Thermoreceptors receive signals from the thermal disturbances and transfer them to the central nervous system via afferent nervous pathways. The receptors can respond not only to temperature, but also much more effectively to temperature change, as shown in Equation (3-1). This means that rapid external cooling or warming may lead to a transient opposite change of internal temperature.

Behavior thermoregulation in human beings is related to conscious thermal sensations and emotional feelings of thermal comfort and discomfort. Behavior thermoregulation to heat and cold modifies the need for autonomic thermoregulatory responses. Hensel summarized various autonomic and behavioral components of temperature regulation, as shown in Table 4-2.

Table 4-2
Components of Thermal Regulation [110, p.15]

Physical factor	Autonomic regulation	Regulatory behavior
Ambient temperature		Migration Seeking sun or shade, etc. Artificial heating and cooling
Heat production	Shivering Non-shivering thermogenesis	Active movements Food intake: Specific dynamic action Warm and cold food
Internal thermal resistance External thermal resistance	Cutaneous blood flow Erection of hair and features Respiration: dry heat loss	Clothing Nest building Seeking shelter Seeking ground surface of various thermal conductivity, wind, water, etc. Air movement by fanning Ventilation
Water evaporation	Sweat secretion Respiration: evaporation heat loss Secretion of nasal and oral glands Moistening of clothing	Moistening of body surface with water, saliva, or nasal fluid
Geometric factor		Body posture Huddling of several individuals

Technical thermoregulation can be considered as an extension of the human regulatory system through technical inventions. The regulation of temperature is shifted from the body to the environment with artificial sensors, controllers, and effectors.

4.4 A Two-node Model of Thermal Regulation

The thermoregulatory system of the human body has been described mathematically by various authors [145–150]. To illustrate the fundamental principles, a simple model developed by Gagge *et al.* is discussed in this section.

In 1971, Gagge *et al.* developed a two-node model for describing the thermoregulatory system of the human body [5]. This model has been further modified and updated [6,144]. As shown in Fig. 4-5, the model assumes that the human body has two concentric shells: skin and core. The skin is represented by a thin shell with mass of m_{sk}, and the body interior by a central core with mass of m_{cr}. The sum of $(m_{sk} + m_{cr})$ is the total body mass (m). Further, the model consists of two systems incorporated in the body, one passive, the other controlling, to regulate body temperature.

The heat balance between the body and its surroundings is described by the passive system, which can be expressed by the following equation:

$$S = M - W - (R + C + E_{dif} + E_{rsw} + E_{comp}) - (E_{res} + C_{res}) \qquad (4\text{-}3)^†$$

---

† For a full explanation of symbols, see *Nomeclature*, pp.125–127

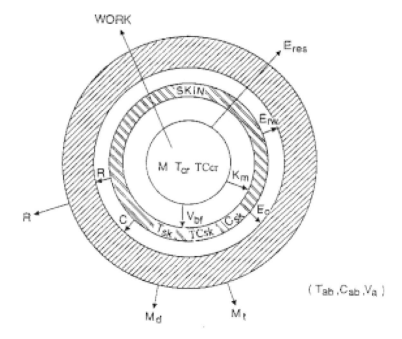

Fig. 4-5   Two-node model of thermoregulation of human body

The heat balance of the body core requires:

$$S_{cr} = M - E_{res} - C_{res} - W - (K_{min} + c_{bl} V_{bl})(T_{cr} - T_{sk}) \qquad (4\text{-}4)^\dagger$$

The heat balance for the skin shell can be described by:

$$S_{sk} = (K_{min} + c_{bl} V_{bl})(T_{cr} - T_{sk}) - E_{sk} - (R + C) \qquad (4\text{-}5)^\dagger$$

$$S = S_{cr} + S_{sk}$$

The rates of change of temperature of the skin, core, and body (°C s⁻¹) can be described as:

$$dT_{sk} = S_{sk} A / TC_{sk} \qquad (4\text{-}6)^\dagger$$

$$dT_{cr} = S_{cr} A / TC_{cr} \qquad (4\text{-}7)^\dagger$$

$$dT_{mb} = \alpha\, dT_{sk} + (1 - \alpha)\, dT_{cr} \qquad (4\text{-}8)^\dagger$$

The controlling system has three mechanisms: the skin blood flow, sweating, and shivering. The skin blood flow is adjusted to change skin temperature for reducing or increasing heat loss, and is assumed to be controlled by temperature signals from the skin and the central core:

$$V_{bl} = (6.3 + 200\ warm_c) / (1.0 + 0.1\ cold_s) \qquad (4\text{-}9)^\dagger$$

---

† For a full explanation of symbols, see *Nomenclature*, pp.125–127

Sweating is an effective mechanism to release extra heat from the body, which is determined by sweating glands that produce the regulatory sweating. The glands are controlled by the mean body temperature signal and the skin temperature signal. The regulatory sweating rate (*regsw*) in $gm^{-2} hr^{-1}$ was identified by Gagge *et al.* [6] as:

$$regsw = 170 \ warm_b \ e^{(warms/10.7)} \qquad (4\text{-}10)$$

The metabolic heat of the body due to shivering can be adjusted to M' by the following equation:

$$M' = M + 19.4 \ cold_s \ cold_c \qquad (4\text{-}11)$$

The variables (*warm_c*, *warm_s*, *cold_s*, and *cold_c*) are functions of the temperature and temperature changes at the skin and the center core, and of the whole body mass described by Equations (4-6) to (4-8). A detailed description of these equations is reported in Gagge *et al.* (1986) [6].

This model has been developed for deriving a standard predictive index of human responses to the thermal environment under isothermal conditions, in which the heat and moisture transfer behavior of clothing is described by the intrinsic insulation of the clothing (*clo* value) and the intrinsic vapor resistance through the clothing. These criteria were developed to describe the heat and moisture transfer behavior of clothing under constant moisture and temperature gradients.

4.5 Dynamic Thermal Interaction Between the Body and Clothing

David (1964) studied the thermal insulation of wool clothing under transient conditions and found that the insulation could increase 50% to 70% above normal due to moisture sorption by the wool [151]. Stuart *et al.* (1989) investigated the heat released by dried wool garments being exposed to a low-temperature and high-humidity environment. They observed that sufficient sorption heat was released during the transients for subjects to perceive the heat as an increase in warmth [152].

De Dear *et al.* (1989) carried out a series of experiments to study the impact of step changes of air humidity on thermal comfort by the use of a thermal manikin and human subjects. From the thermal manikin experiments they found that 37–42% of the heat involved during absorption or desorption of moisture by wool garments resulting from the humidity change influenced the sensible heat balance of the wearer. From the measurement of temperatures at the skin surface they observed significant changes in skin temperature, especially when they were wearing wool garments [153]. Therefore, Gagge's model is valid for describing the thermal comfort state of a clothed man under steady environmental conditions, but is not valid for transient conditions.

Shitzer and Chato (1985) studied the heat and mass transfer of the clothing–air–skin system. They considered the heat and mass transfer in the system as a steady-state problem in a one-dimensional model composed of five layers (ambient air, fabric, airspace, skin, and body core). For the body, their model assumed constant physical properties, temperature-dependent skin thermal conductivity, and no energy penetration to the body core. For the fabric, their model considered the fibers as always in equilibrium with the adjacent air. This work represented a significant development in the study of simultaneous heat and mass transport through the skin–fabric system [154].

In an attempt to describe this dynamic behavior, Jones *et al.* (1990) reported a model of the transient response of clothing systems that took account of the sorption behavior of fibers, with the assumption that the fibers are always in equilibrium with the surrounding air. They compared the prediction of heat loss by the model with experimental data from thermal manikin tests and found reasonable agreement [155]. Further, they combined this clothing model with Gagge's two-node model to investigate the interactions between the body and clothing [156]. Data to confirm the validity of this combined model has not yet been reported.

Li and Holcombe (1998) [157] developed a mathematical model by interfacing the model for a naked body (Equations (4-3) to (4-11)) with a heat and moisture transfer model of a fabric in which the complex moisture sorption processes were taken into account (see Sections 5.3–5.9). The boundary conditions between the skin and clothing can be described by the following equations:

heat:
$$M_t = h_{ti}(T_{sk} - T_{fi})$$
(4-12)[†]

mass:
$$M_d = h_{ci}(C_{sk} - C_{fi}) + L_{sk}\frac{\partial C_{sk}}{\partial t}$$
(4-13)[†]

where M_t is the heat flow from the skin ($= p_h(R + C)$) and M_d is the moisture flow from the skin ($= p_m E_{sk}/\lambda$).

This model has been used to describe, mathematically, the dynamic heat and moisture transfer behavior of the body–clothing–environment system under transient conditions. With specification of the physical activity and ambient conditions, the model is able to predict the thermoregulatory responses of the body, together with the temperature and moisture profiles in the clothing. Li and Holcombe (1998) reported a series of experimental measurements with garments made from fibers with different levels of hygroscopicity [157]. The experimental results were compared with predictions by the model for an ambient condition with changing humidity.

The measured and predicted temperature profiles in the clothing and on the skin when wearing wool garments is shown in Fig. 4-6. A small change in core temperature and greater changes in skin and fabric temperature were predicted by the model. Comparing with the experimental results, the predictions agreed well with the measurements, though the measured temperatures showed greater variations than those predicted.

Fig. 4-7 shows the predictions of the changes in the relative humidity of the microclimate. Comparing with the experimental measurements, the predicted relative humidities showed good agreement with the measured ones. Similar changing trends were observed in both the predictions and experimental results, though the measured relative humidity in the microclimate appeared lower than that predicted.

Theoretical predictions and physiological measurements were carried out also for a polyester garment (see Figures 8-1 and 8-2 in Section 8). Essentially, good agreement in temperature and relative humidity profiles was observed between theoretical predictions and experimental results. These results showed that the model is able to predict the transient heat and moisture transport behavior of garments made from highly and weakly hygroscopic fibers in dynamic wear situations. With good prediction of the corresponding dynamic skin temperature responses of the body, the model can also be used to show how the transient behavior of clothing interacts with the skin and influences the thermoregulatory responses of the body. Further, the model has been used to predict the dynamic heat and moisture transfer of clothing and skin temperature changes when human subjects wear highly and weakly hygroscopic fibers and are exposed to large environmental changes such as rain and walking from indoor to outdoor ambient conditions. The theoretical

---

† For a full explanation of symbols, see *Nomenclature*, pp.125–127

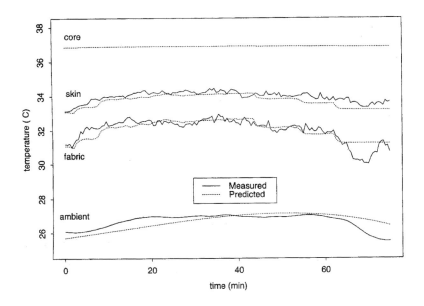

Fig. 4-6   Temperature profile in the clothing–body system [157]

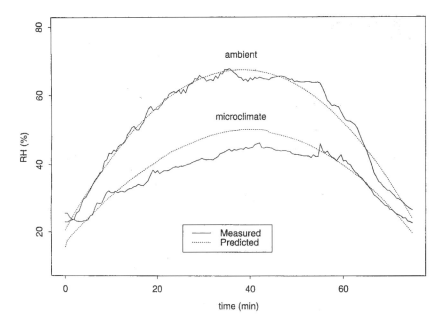

Fig. 4-7   Relative humidity changes in the clothing microclimate [157]

predictions that were made before the actual experiment were confirmed by physiological measurements during wear trials [158].

5. DYNAMIC HEAT AND MOISTURE TRANSFER IN TEXTILES

5.1 Physical Processes Involved in Clothing Comfort

In the previous sections, the psychological, neurophysiological, and thermal physiological aspects of comfort have been discussed. The psychological perceptions are formulated on the basis of the neurophysiological signals that are derived from various sensory nerve endings. Thermophysiological regulatory responses of the body are also triggered by the signals from thermal receptors located throughout the body. The signals from the nerve endings are generated essentially by various physical stimuli from external environments, especially from clothing that covers most of our body.

The physical processes that generate those stimuli include heat transfer by conduction, convection and radiation, moisture transfer by diffusion, sorption, wicking and evaporation, and mechanical interactions in the form of pressure, friction, and dynamic irregular contact. Comprehensive research has been carried out to study the mechanisms of heat and moisture transfer processes in clothing in the past few decades. However, the mechanisms of the mechanical interaction between clothing and the body have not been investigated so thoroughly. This section will focus on the heat and moisture transfer processes. Section 6 will discuss the mechanisms of various thermal and moisture sensations during wear. Section 7 will review progress on research of the mechanical interactions and their influence on the comfort perceptions.

5.2 Heat and Moisture Transfer in Clothing

The heat and moisture transfer behavior of clothing has long been recognized to be critically important for human survival. A great deal of work has been carried in this area. In 1970, Fourt and Hollies [2] did a comprehensive survey on the literature concerning clothing comfort and function, with special emphasis on thermal comfort. Seven years later, Slater [159] carried out an extensive review on the comfort properties of textiles, including the measurement of their thermal resistance, water vapor transmission, liquid-moisture transmission, and air permeability.

Hollies and Goldman (1977) [3] reviewed the criteria in evaluating thermal comfort performance of clothing. They used a number of equations to describe the heat and moisture transfer in clothing:

$$\text{Convective heat loss, } H_c = k_c \cdot A \cdot (T_{sk} - T_{ab}) \tag{5-1}$$

$$\text{Evaporative heat loss, } H_e = k_e \cdot A \cdot (P_{sk} - P_{ab}) \tag{5-2}$$

$$\text{Mean radiant temperature, } MRT = 1 + 2.22 \sqrt{V} \cdot (T_g - T_{ab}) + T_{ab} \tag{5-3}$$

$$\text{Adjusted dry bulb temperature, } AT_{db} = (T_{ab} + MRT)/2 \tag{5-4}$$

where k_c is a coefficient for convective heat transfer, which involves not only the still air layer around the body but also the thermal characteristics of the clothing worn; A is the surface area of the body; T_{sk} is the mean weighted skin temperature of the surface of the body; T_{ab} is the dry bulb temperature; k_e is the evaporative coefficient, as determined by the Lewis relationship ($k_e = 2.2\,k_c$); P_{sk} is the saturated vapor pressure of water at skin temperature; P_{ab} is the ambient vapor pressure; V is the wind velocity; T_g is the globe temperature; AT_{db} is a combined coefficient for clothing thermal insulation, incorporating both convective and radiative heat transfers.

The most widely used unit in the US is the *clo*, which was proposed by Gagge and his colleagues at the Pierce Foundation in 1941 [161]. One *clo* was defined as the intrinsic insulation of the typical business suit worn in those days (see Fig. 4-3) and equals $0.155°C\ m^2\ W^{-1}$.

Mecheels and Umbach (1977) [162] reviewed the psychrometric range of clothing systems. They pointed out that the thermal properties of a clothing system are determined by its resistance to heat transfer, R_c, and its resistance to moisture transfer, R_e, where:

$$R_c = A\,(T_s - T_a)/H_c \text{ in } 155^{\circ}\text{C} \cdot \text{m}^2 \cdot \text{W}^{-1} \qquad (5\text{-}5)$$

$$R_e = A_x\,(P_s - P_a)/H_e \text{ in } 155 \text{ mm Hg} \cdot \text{m}^2 \cdot \text{W}^{-1} \qquad (5\text{-}6)$$

and

$$i_m = 0.45\,(R_c \text{ clothing}/R_e \text{ clothing})$$

where i_m is the moisture permeability index proposed by Woodcock in 1962 [160]. When the clothing is a perfect vapor barrier, $i_m = 0$; when a clothed and 100% sweat-wetted man achieves the full evaporative potential of a ventilated wet-bulb thermometer, $i_m = 1$.

They pointed out that, through these two values of resistance, the minimum ambient temperature and the maximum ambient temperature could be determined. The minimum ambient temperature was defined as the temperature at which the thermoregulation system of the human body is within the cold range, i.e. the moisture concentration near the skin is close to that of the environment so that the moisture flow can be neglected. The maximum ambient temperature is the temperature at which the human thermoregulation system reaches the upper temperature limit range, where the wearer of a clothing system must prevent his core temperature, under certain comfort conditions, from rising by making use of evaporative cooling. The difference between the maximum and minimum ambient temperatures is called the psychrometric range of the clothing system.

The resistance to heat and moisture transfer and the psychrometric range can be measured by using a thermal manikin and a skin model that were developed by the Hohenstein Institute. The parameters are dependent on the clothing design and the way it is worn, the textile material, and the wind velocity.

Breckenridge (1977) [163] surveyed the literature on the effects of body motion on convective and evaporative heat exchanges of clothing. The thermal insulation of clothing is dependent on a number of factors: thickness and number of layers, fit, drape, fiber density, flexibility of layers, and adequacy of closures. Thermal insulation values and the moisture permeability index (i_m) of military clothing assemblies were routinely measured by using a standing, life-sized copper manikin in the Military Ergonomics Division at USARIEM. A sensitive balance was used to monitor a subject's weight loss during activity for the estimation of evaporative heat loss. For measuring the 'pumping' coefficients associated with body motion, Mecheels (1971) [164] reported a walking manikin at the Hohenstein Institute, and Madsen and Fanger (1975) developed a manikin that could simulate pedaling motions on a bicycle, as well as stand erect [163].

All these research publications regarded the heat and moisture transport processes as two independent processes, which is largely applicable for the wear situations under various steady states. During humidity transients, the heat and moisture transport processes are coupled. The thermal insulation of clothing is influenced by the moisture sorption of the textile fibers. Therefore, the measurement methods and criteria may not be appropriate for the evaluation of the thermal comfort of clothing under dynamic wear conditions. In the last decade, the dynamic heat and moisture transport behavior of clothing and their influence on thermal and moisture perceptions have become the main focus of research in the field. The next section reviews the major progress in this area.

5.3 Dynamic Heat and Moisture Transfer in Fabric

The coupled heat and moisture transfer in textile fabrics has been widely recognized as being very important for understanding the dynamic thermal comfort of clothing during wear. In 1939, Henry proposed a mechanism for the transient diffusion of moisture and heat into an assembly of textile fibers [165], and he further described a model in 1948 [166]. In the model, Henry developed a system of differential equations to describe the processes involved. Two of the equations involve conservation of mass and energy. The third equation relates fiber moisture content to the adjacent air.

As shown in Fig. 5-1, in a small element of fabric of unit area and thickness, packed with fibers exposed to a moisture gradient and a temperature gradient, water vapor diffuses through the interfiber spaces to be absorbed or desorbed by the fibers. To simplify a mathematical description of the process, a number of assumptions were made: (i) The volume changes of the fibers due to changing moisture content can be neglected; (ii) Moisture transport through fibers can be ignored as the diffusion coefficient of water through fibers is negligible compared with that through air; (iii) The orientations of fibers in the fabric plays a minimum role in the water vapor transport process as the diameters of fibers are small and water vapor can travel much more rapidly in the air than in the fibers; (iv) Instantaneous thermal equilibrium between the fibers and the gas in the interfiber space is achieved during the process, as most textile fibers are of very small diameter and have a very large surface/volume ratio.

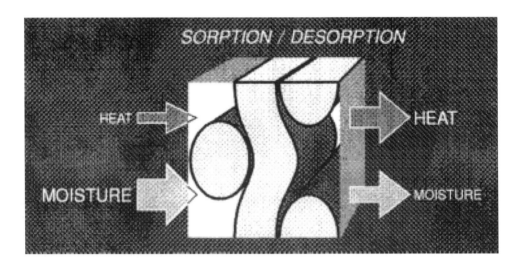

Fig. 5-1   Coupled heat and moisture transfer in a fabric

On the basis of these assumptions, a mass balance equation which considers the moisture accumulation by both the air and the fibers and the moisture transport through the air space, can be written as:

$$\varepsilon \frac{\partial C_a}{\partial t} + (1-\varepsilon)\frac{\partial C_f}{\partial t} = \frac{D_a \varepsilon}{\tau}\frac{\partial^2 C_a}{\partial x^2} \tag{5-7}$$

In this equation, the first term on the left-hand side describes the accumulation of water vapor in the interfiber space, and the second term describes the accumulation of absorbed water in the fibers. The moisture transport through the interfiber air space is described by the term on the right-hand side.

A second equation for the conservation of heat energy can be derived by considering changes in the heat content of the volume element that arise from a number of processes: conduction into or out of the element, change in phase of the water vapor (sorption or desorption), and temperature changes of the fibers and of the air in the interfiber space. The equation for energy conservation can be written as:

$$C_v \frac{\partial T}{\partial t} - \lambda(1-\varepsilon)\frac{\partial C_f}{\partial t} = K\frac{\partial^2 T}{\partial x^2} \qquad (5\text{-}8)$$

In this equation, C_v and λ are dependent on the concentration of water absorbed by the fibers. These two equations are not linear and contain three unknowns (C_f, T, and C_a). A third equation that describes the relationship between C_a and C_f, i.e. the water exchange between the fiber and its surrounding air, is needed in order to progress to a solution.

5.4 Moisture Exchange Between Fiber and Air
5.4.1 The Drying Behavior of Fabrics
The moisture exchange between a fiber and its adjacent air is a complex process, depending on whether the moisture is present as liquid on the fiber surface, or as vapor stored internally. This is best illustrated from the drying behavior of fabrics.

Lyons and Vollers [167] analyzed the drying process of textile materials. They found that it has three distinct stages. In the first stage, a wet fabric adjusts its temperature and moisture flows with its surrounding environment. The second stage is a 'constant drying rate' period, in which the drying rate remains constant as the rates of heat transfer and vaporization reach equilibrium. Liquid moisture moves within the fabric to maintain a saturation condition at the surface. The third stage is a 'declined drying rate', during which moisture flow to the surface is insufficient to maintain saturation and the plane of evaporation moves into the fabric. Fibers begin to desorb moisture until equilibrium is reached between the fabric and the environment.

Fig. 5-2 illustrates the drying behavior of wool and polyester fabrics at 25°C and 25% r.h., as a function of time [121]. The vertical axis is expressed as percentage excess moisture, i.e. the water content above the equilibrium regain. When fabric water content is above the fiber saturation moisture content, the drying rates of both fabrics are constant and approximately the same, because the drying process is determined by a surface evaporation process. When fabric water content decreases to below the saturation moisture content, the drying rate declines as liquid water at the fiber surface has evaporated and the water absorbed within the fibers is released. This drying process continues until equilibrium with the ambient conditions is reached. The difference between the wool and polyester fabrics is that the constant rate period by evaporation is prolonged for the polyester as its saturation moisture content is below 1%. Meanwhile, the 'declined period' is prolonged with wool, as wool has a much higher saturation water content (up to 36%).

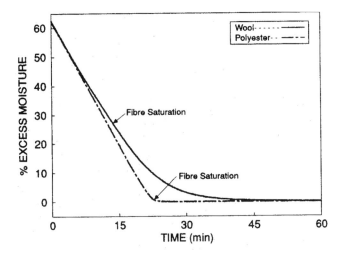

Fig. 5-2   Change in water content of moist fabrics during drying [121]

The corresponding fabric temperature during the drying process is shown in Fig. 5-3 [121]. When the fabrics' water content is above their saturation regain, the temperatures of both fabrics are approximately the same and below the ambient temperature, because the dominant process is evaporation of free water. As their water contents approach their equilibrium regain, their temperatures begin to rise until all excess moisture has evaporated and equilibrium is achieved with the surroundings. The temperature change of the wool fabric behaves distinctly differently from that of the polyester fabric during the drying process. The wool fabric shows a longer transition in temperature from wet to dry than the polyester fabric. This reflects the greater moisture sorption capacity of wool and its influence on the heat and moisture exchange between fabric and environment.

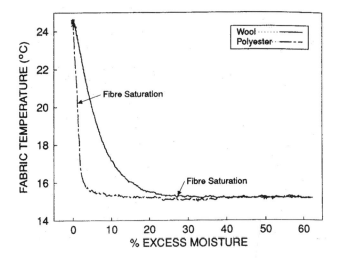

Fig. 5-3   Relationship between fabric temperature and excess moisture [121]

These observations suggest that two separate equations need to be developed to describe the moisture exchange between fiber and surrounding air, which are discussed in the following two sections. The liquid transport in a fabric influences the mass transfer (i.e. conservation equation) but does not involve energy exchange. However, it determines the dynamic distribution of liquid in the fabric, which in turn determines the front of water evaporation. Experimental investigations have shown that the liquid transfer has significant impact on the heat transport processes in clothing, and on its thermal comfort and tactile comfort performance [168]. Gibson (1994) [169] developed a set of complex mathematical equations to describe the multiphase heat and mass transfer in hygroscopic porous media with applications to clothing materials. However, solutions for the equations have not been reported.

5.4.2 Evaporation and Condensation

Crank (1975) [170] described the evaporation – condensation process mathematically, as shown in Equation (5-9). This equation applies when fabric water content is above the saturation regain of the fiber; that is, liquid water is present in capillaries within the fabric structure or at the fiber surface. The exchange of water is an evaporation – condensation process.

$$\frac{\partial C_f}{\partial t} = h_{cf} S_v \left(C_{fs} - C_a \right) \tag{5-9}$$

where C_{fs} is the water concentration in the fiber surface (kg m^{-3}); C_a is the water concentration in the adjacent air (kg m^{-3}); h_{cf} is the mass transfer coefficient at the fiber surface (ms^{-1}); and S_v is the specific volume of the fabric (m^{-1}).

5.4.3 Moisture Sorption and Desorption

When the fabric water content is below the saturation regain of the fiber, the exchange of water can be considered as a sorption or desorption process. David and Nordon (1969) [171] developed an experimental relationship between the rate of change of water content of the fibers and the absolute difference between the relative humidity of the air and fiber. The rate equation was given as:

$$\frac{1}{\varepsilon} \frac{\partial C_f}{\partial t} = \left(H_a - H_f \right) \chi \tag{5-10}$$

where

$$\chi = k_1 (1 - \exp[k_2 \, | \, H_a - H_f |]) \tag{5-11}$$

and k_1 and k_2 are parameters that are adjustable according to experimental results.

David and Nordon incorporated several features omitted by Henry and developed a solution of the equations by finite difference methods, which provided space–time relationships for moisture concentration and temperature within the air–fiber mass. The authors stated that the model did not consider the sorption–desorption kinetics of the fibers, and that proper boundary conditions need to be specified in evaluating the coupled heat and moisture transfer processes in clothing during wear.

In 1986, Farnworth [172] reported a numerical model of the combined heat and water vapor transport in clothing, in which the mass of absorbed water was assumed to be directly proportional to the relative humidity; also, the three forms of water (vapor, liquid, and absorbed water) were in equilibrium with each other locally. Therefore, the model did not take into account the complexity of the moisture sorption isotherm and the sorption kinetics of textile fibers.

Wehner (1987) [173] investigated the influence of moisture sorption by fibers on the moisture flux through the air spaces of a fabric. He developed two mathematical models to describe the processes. In the first model, diffusion within the fiber is considered to be rapid and the fiber moisture content is always in equilibrium with the air at the fiber surface. Hence, the dominant mass transfer resistance for the sorption process is assumed to be the diffusion of water molecules through the air to the fiber surface. In the second model, the sorption kinetics of the fiber are assumed to be Fickian diffusion and the dominant mass transfer resistance is molecular diffusion of water molecules within the fiber interior. Therefore, the fiber moisture content lags behind the changes in the moisture content of the air at the fiber surface. In these models, the interaction between moisture absorption and heat of sorption is ignored.

Downes and Mackay (1958) [174] and Watt (1960) [175,176] studied the kinetics of the uptake of water vapor by wool and found it to be a two-stage process. The first stage obeys Fick's laws of diffusion, with a concentration-dependent diffusion coefficient until absorption to quasi-equilibrium. The second stage is much slower than the first and is accompanied by structural changes within the fibers. The relative contributions of the two stages to the total uptake depend on the initial regain of the fiber and the stage of absorption.

On the basis of these findings, Li and Holcombe (1992) [177] assumed that the water vapor uptake rate of the fiber consists of two components associated with the two stages of sorption identified. The first stage is represented by Fickian diffusion, and the second stage sorption follows an exponential relationship. Thus, the water exchange equation can be written as:

$$\frac{\partial C_f}{\partial t} = (1-p)R_1 + pR_2 \qquad (5\text{-}12)$$

where R_1 and R_2 are the first-stage sorption rate and the second-stage sorption rate respectively, and p is the proportion of uptake occurring during the second stage.

The first-stage sorption rate, R_1, can be obtained by considering the sorption–desorption process as Fickian diffusion. Crank [170] has shown that the radial diffusion of moisture in a cylindrical medium is governed by the following relationship:

$$R_1 = \frac{\partial C_f}{\partial t} = \frac{1}{r}\frac{\partial \left(rD_f \partial C_f\right)}{\partial r} \qquad (5\text{-}13)$$

In this model, the moisture content at the fiber surface is assumed to be in instantaneous equilibrium with the moisture content of the adjacent air. Hence:

$$C(x, R_f, t) = f\{C_a(x,t)\} = \rho W_c(H_f)$$

where R_f is the mean radius of the fibers (m); W_c is the fractional water content at the fiber surface; H_f is the fractional relative humidity of the adjacent air; and ρ is the density of the fibers (kg/m³). The relationship between W_c and H_f can be determined from the sorption isotherms of textile fibers.

According to the experimental data presented by Watt [175,176], Li and Holcombe [177] specified the proportion of uptake (p) during the second stage as:

$$p = 0.0 \text{ when } W_c < 0.185, \; t < t_q \tag{5-14}$$

$$p = 0.5 \text{ when } W_c \geq 0.185, \; t < t_q \tag{5-15}$$

$$p = 1.0 \text{ when } t \geq t_q \tag{5-16}$$

where t_q is the time to reach quasi-equilibrium ($t_q \approx 540$ seconds, according to Watt).

The second stage sorption rate (R_2) relates the local temperature, humidity, and sorption history of the fiber at each point in the fabric, which was assumed as the following form:

$$R_2 = s_1 \text{ sign}(H_a - H_f) \exp(s_2 / |H_a - H_f|) \tag{5-17}$$

where s_1 and s_2 are constants that may be determined by experiment.

5.5 Boundary Conditions

To solve equations (5-7), (5-8), (5-9), (5-12) and (5-13), the initial condition and boundary conditions must be specified. The initial condition is determined by the history of the thermal and moisture environment of the fabric. For instance, when a fabric is equilibrated in a given environment, the temperature and moisture content can be regarded as uniform throughout the fabric at known values:

$$T(x,0) = T_0 \tag{5-18}$$

$$C_a(x,0) = C_{a0} \tag{5-19}$$

$$C_f(x,0) = f(H_{a0}, T_0) \tag{5-20}$$

David and Nordon (1969)[171,178] studied the situation where the fabric boundaries are exposed to an air stream of new moisture content C_{ab} and temperature T_b. By assuming the rate of moisture diffusion and heating are sufficiently rapid that the bulk moisture content and the temperature of the air stream are equal to those at the surface of the fabric, they specified the pertinent boundary conditions as:

$$T(0,t) = T(L,t) = T_b \tag{5-21}$$

$$C_a(0,t) = C_a(L,t) = C_{ab} \tag{5-22}$$

Nordon and David numerically solved Equations (5-7), (5-8), (5-10), and (5-11) with boundary conditions (5-21), (5-22) by finite difference method. Those boundary conditions cannot be achieved in practice because a boundary layer of air exists and limits the transfer of heat and moisture between the fabric and the air stream. Li and Holcombe [177] used a set of equations to describe this layer of air, which takes account of the convective nature of the boundary conditions:

$$D_a \frac{\partial C_a}{\partial x}\bigg|_{x=0} = h_c(C_a - C_{ab}) \tag{5-23}$$

$$D_a \frac{\partial C_a}{\partial x}\bigg|_{x=0} = -h_c(C_a - C_{ab}) \tag{5-24}$$

$$K\frac{\partial T}{\partial x}\bigg|_{x=0} = h_t\left(T_a - T_b\right) \tag{5-25}$$

$$K\frac{\partial T}{\partial x}\bigg|_{x=0} = -h_t\left(T_a - T_b\right) \tag{5-26}$$

In simulating the heat and moisture processes during the fabric–skin contact, Li *et al.* used various equations to specify the boundary conditions between fabric and skin in investigating fabric coolness to the touch [179], fabric dampness perception [115], moisture buffering behavior of hygroscopic clothing [32], and the interaction between thermoregulatory responses and hygroscopic clothing during humidity transients [158].

5.6 Physical Properties of Fibers and Fabrics

To solve equations (5-7), (5-8), (5-9), (5-12), (5-15), (5-18), (5-19), (5-20), (5-23), (5-24), (5-25), (5-26), a number of numerical values of relevant fiber and fabric properties need to be specified. In Table 5-1, the numerical values for the thermal and moisture sorption characteristics of fibers and fabrics are summarized on the basis of previous research work. David and Nordon (1969) [171] reported the relationship between fiber moisture content and the relative humidity at the fiber surface for wool. Those for cotton and polyester were determined from the data published by Rae and Rollo [180] and Urguhart [181]. Rae and Rollo also reported the relationship between heat of sorption and fiber moisture content [180]. Schneider [182] studied the relationship between fabric thermal conductivity and moisture content, the results of which were used to obtain the equations in Table 5-1. The relationship between fabric volumetric heat capacity and fiber moisture content was obtained by proportion, based on the specific heat of the dry fibers and that of water (4.184 kJ/kgK [183]). For the wool/polyester blend fabric, the polyester was treated as non-absorbent, and each of these variables and relationships was determined for a fabric containing wool present in the proportion represented in the blend. The value of the diffusion coefficient of water vapor in air used was 2.49×10^{-5} (m/s) [183].

Table 5-1
Numerical Values used in the Theoretical Calculations [115]

Fiber	Wool	Cotton	Polyester
Fiber moisture content W_c	-0.0006176 {ln[0.02293 $-\ln(H_f + 0.0083)] -1.5713$}	-0.03729 {ln[0.0939 $-\ln(H_f + 0.0946)] -0.9783$}	$0.006244\ H_f$
Mean fiber radius R_f (μm)	9.7	6.0	10.1
Heat of sorption (kJ kg⁻¹)	$1602.5\exp(-11.727\,W_c)$ + 2522.0	$1030.9\exp(-22.39\,W_c)$ + 2522.0	2522.0
Fabric thermal conductivity k (mW mK⁻¹)	$29.4 + 27\ W_c$	$44.1 + 63\ W_c$	$40.4 + 23\ W_c$
Diffusion coefficient of water in the fiber D_f (m² s⁻¹)	5.8×10^{-13}	2.5×10^{-13}	3.9×10^{-13}
Specific heat of dry fibers (J kg⁻¹ K⁻¹)	1338	1380	1255
Fiber density (kg m⁻³)	1310	1350	1220
Fabric volumetric heat capacity C_v (kJ m⁻³ K⁻¹)	$((1752 + 4184\ W_c) / (100 + W_c)$	$(1863 + 4184\ W_c) / (100 + W_c)$	$(1531 + 4184\ W_c) / (100 + W_c)$

5.7 Method of Solution

5.7.1 Moisture Diffusion into the Fiber

To solve equations (5-7) to (5-26), the process of moisture diffusion into the fibers needs to be solved. Details of the method used to determine R_1 are beyond the scope of this monograph and a full description was reported by Li and Holcombe in 1991 [184]. Briefly, moisture is considered diffusing radially into a cylindrical medium (fiber) with a constant diffusion coefficient (D_f), and at the fiber surface the moisture content is in equilibrium with the moisture content of the adjacent air.

Initially, the moisture content is uniform, and the boundary condition at the center of the fiber is one of symmetry:

$$C_f(x, r, 0) = C_0 \tag{5-27}$$

and

$$\frac{\partial C_f}{\partial r}\bigg|_{r=0} = 0 \tag{5-28}$$

The moisture content at the fiber surface ($r = R_f$) is in equilibrium with the moisture content of the adjacent air, so that:

$$C_f(x, R_f, 0) = f(C_a(x,t)) = \Phi(x, t) \tag{5-29}$$

where f is the moisture sorption isotherm of the fiber. An analytical solution to Equation (5-13) was derived by using Crank's solution [170]. This approach was demonstrated to be able to describe the dynamic heat and moisture transfer of wool fabric with good accuracy by comparing with experimental results. However, Crank's solution is truncated, which leads to the corresponding algorithm that needs a strict restriction to time step. Therefore, long computation time is required to achieve acceptable accuracy.

Li and Luo (1998) [185] improved the model further by applying a direct numerical solution to Equation (5-12) without using Crank's solution. More importantly, the two-stage moisture sorption of wool is described uniformly by Equation (5-12). The first stage is considered as a Fickian diffusion process with a moisture-dependent diffusion coefficient. The second stage is also expressed as a Fickian diffusion equation with a diffusion coefficient that is time-dependent and corresponds to the nature of the structural changes during moisture sorption in the fibers. This model is able to clearly illustrate the moisture and heat transport processes by two-dimensional space and time diagrams.

5.7.2 Numerical Solution to the Main Equations

In 1967, Nordon and David reported a numerical solution to Equations (5-7), (5-8), (5-10), and (5-11) of the their model by using the Crank–Nicholson implicit finite difference technique [178]. Li and Holcombe [184] also used finite difference techniques to obtain solution for Equations (5-7) to (5-9), (5-12), (5-13), and (5-17) of the two-stage model. With the solutions, the profiles and waves of temperature and moisture transfer can be obtained. The model can be applied to either a sorption–desorption process or an evaporation–condensation process, depending on the initial and the new boundary conditions. The calculations are performed stepwise in small time intervals to follow the development of T, C_a, and C_f.

5.8 Moisture Sorption of Wool Fabrics

Li and Holcombe [177] studied the dynamic heat and moisture transport processes during moisture sorption by setting-up an experimental apparatus, as shown in Fig. 5-4. In the experiment, fabrics were equilibrated in a cell controlled at 20°C and 0% r.h. Then, the r.h. in the cell was changed to 99%. The water content changes during sorption were obtained by weighing the fabric continuously. The temperature changes in the fabric were also recorded by inserting thermocouple wires into the surface of fabric samples.

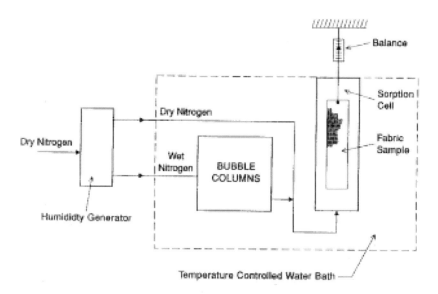

Fig. 5-4   Schematical diagram of the experimental apparatus [177]

The experimental results of both temperature and water-content change were compared with the predictions from three models: the Nordon–David Model, the Fickian diffusion model, and the two-stage sorption model. As shown in Fig. 5-5 [177], the Nordon–David model can predict the trend of the water-vapor uptake in general, but does not fit the experimental observations very well. The prediction from the Fickian diffusion model follows the first 540 seconds of actual sorption closely, but its predicted subsequent rate of water uptake is too high in comparison with the experimental observation. This result agrees with Watt's findings that, for single wool fibers, the first-stage absorption to a quasi-equilibrium obeys Fick's laws of diffusion and that the time to reach the quasi-equilibrium is about 540 seconds. Fig. 5-5 illustrates that the two-stage sorption model has the best agreement with the experimental observation overall.

The authors also compared the observed simultaneous temperature changes with the predictions from the model, as shown in Fig. 5-6. The simultaneous fabric surface temperature changes were the result of heat released during the water-vapor sorption by the fibers in the fabric. All the temperature curves predicted by the three models showed similar trends to the experimental curve: a very rapid initial temperature rise at the fabric surface, resulting from small increase in water vapor uptake, followed by gradual decrease associated with the continuing water vapor uptake. The predicted temperature change by the two-stage model showed the best fit with the experimental observations.

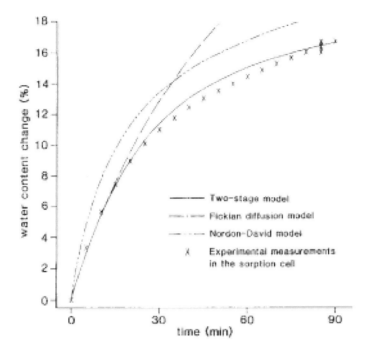

Fig. 5-5   Moisture uptake of a wool fabric during humidity transients [177]

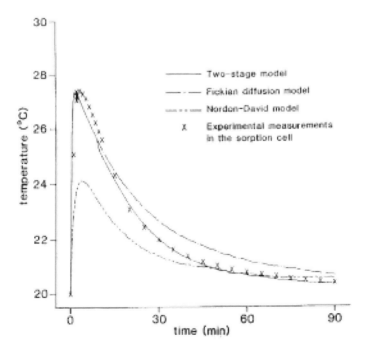

Fig. 5-6   Temperature change at the surface of a wool fabric during humidity transient [177]

5.9 Behavior of Fabrics made from Different Fibers

Further work was carried out by Li and Luo [186], to investigate the dynamic moisture diffusion into hygroscopic fabrics made from different fibers. Four fabrics, made from wool, cotton, porous acrylic, and polypropylene, were tested by using a similar experimental set-up to Fig 5-4. The differences among fabrics with different levels of hygroscopicity in dynamic moisture transfer mechanisms were investigated by using different mathematical models to describe the coupled heat and moisture transfer in the fabrics. The experimental results and theoretical predictions were then compared. The characteristics of the fabrics examined are shown in Table 5-2.

Table 5-2
Basic Characteristics of Fabric Samples

Fiber type	Fiber diameter (mm)	Yarn count (tex)	Fabric structure	Weight (gm⁻²)	Thickness (mm)
Wool	20.6	20.4	double jersey	272	2.96
Cotton	13.3	19.7	double jersey	275	2.19
Porous acrylic	18.4	21.3	double jersey	287	2.14
Polypropylene	20.0	18.3	double jersey	279	2.42

Fig. 5-7 shows the water vapor uptake of the fabrics during sorption from 0% r.h. to 99% r.h. in a single step. The wool fabric had significantly greater water vapor sorption in total than the other fabrics. It also had the highest initial sorption rate, followed by cotton, porous acrylic, and polypropylene fabrics. Differences in water vapor uptake between fabrics increased with sorption time and were in the order of their respective levels of hygroscopicity.

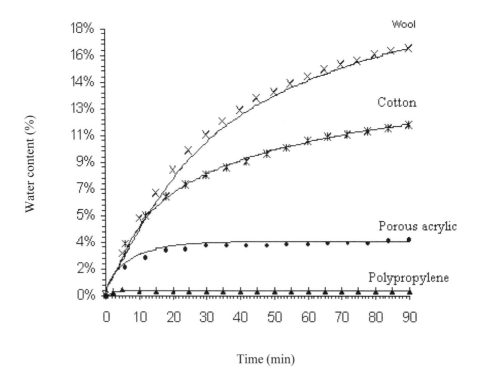

Fig. 5-7   Moisture uptake of fabrics made from various fibers during humidity transients

Fig. 5-8 shows the temperature changes at the surface of the test fabrics during sorption of water vapor. Wool showed the highest initial temperature increase, followed by cotton, porous acrylic, and polypropylene.

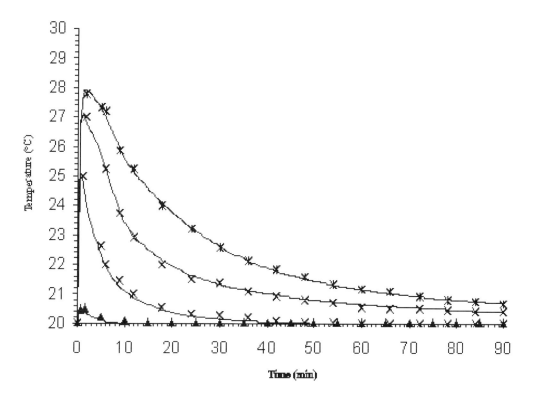

Fig. 5-8   Temperature changes at the surfaces of fabric made from various fibers during humidity transients

Theoretical predictions of water vapor uptake and surface temperature change during the dynamic moisture transfer processes for the fabrics used in the experiment are also presented in these two figures. The features of the experimental results are described well by the models. The two-stage sorption model gives a good description of the water vapor transfer process in highly hygroscopic fabrics such as wool and cotton, and the Fickian diffusion model describes the water vapor transfer in weakly hygroscopic fabrics such as acrylic and polypropylene. This is a clear indication of the differences in sorption mechanisms between highly hygroscopic and weakly hygroscopic fabrics.

These results demonstrate that strongly hygroscopic and weakly hygroscopic fabrics are significantly different in dynamic moisture transfer behavior during environmental moisture transients. Highly hygroscopic fabrics such as wool and cotton show greater mass and energy exchange with the environment than weakly hygroscopic fabrics such as porous acrylic and polypropylene.

6. PHYSICAL MECHANISMS OF TEMPERATURE AND MOISTURE SENSATIONS

6.1 Transient Temperature and Moisture Sensations

In modern living conditions, many workplaces and modes of transport are climate controlled, so the demand for warm clothing has been substantially reduced. As a result, most of our clothing is often in contact with the skin. Consumers are now more conscious of the sensory perceptions against the garments that they are wearing. In summer, the perception of coolness during the dynamic momentary contact is a favorable sensation demanded by wearers. In Chinese history, cotton, linen, and silk fabrics were, a long time ago, recognized and used as comfortable materials for summer wear, due to their coolness and light weight. In an attempt to expand the market for wool and change the 'warm fiber' image, the International Wool Secretariat has, in recent years, produced and promoted lightweight wool fabrics (called 'cool wool'), that are made of tight-twist type yarns.

In winter, the warmth to the touch of textiles provides comfort and favorable sensations to wearers. In wet weather, wearers like to feel dry and warm. During exercise, they like the garment to 'breath' (i.e. not be clammy), and be able to transfer the extra heat away to reduce heat stress. When exposed to sudden environmental changes, they like the garment to protect them against these. All these thermal and moisture-related sensations contribute to the overall perception of comfort experienced during wear.

6.2 Coolness to the Touch

Fabric coolness to the touch is a skin sensation that is related to the transient heat and moisture transfer between fabric and skin. This sensation has a significant impact on the perception of comfort in warm and hot environments. Schneider *et al.* (1990) [39] and (1996) [40] reported that a smooth, lightweight, wool fabric was perceived to be cooler than less hygroscopic fabrics that were matched in construction, such as polyester, during contact with the skin in warm, humid environments. In the investigations, pairs of fabrics, matched in construction and surface features but made from different types of fibers, were evaluated subjectively. The fabric samples were placed across the skin of the inner forearm of test subjects. Evaluation took place at ambient temperatures of 20°C and 28°C, and r.h. in the range from 10% to 90%. A typical result is shown in Fig. 6-1. The wool fabric was consistently rated cooler than the polyester fabric for all the climate conditions tested.

The authors also observed that the amount of moisture desorbed from the wool fabric was significantly higher than that from the polyester fabric, and the skin temperature decreased faster and recovered more slowly after contact with the wool fabric compared with the polyester fabric. They proposed a mechanism by which the coolness to the touch of smooth, lightweight fabrics was enhanced by fiber hygroscopicity, due to desorption of a very small quantity of water from the fibers.

Li *et al.* (1993) [179] applied the coupled heat and moisture transfer model that was discussed in Section 5, to describe the dynamic heat and moisture exchange between the skin and fabric during contact. An experiment was carried out to measure the surface temperatures of the skin and fabric during contact, by using thermocouples. The measured fabric temperatures for wool and polyester fabrics were compared with the predicted temperatures from the model, as shown in Fig. 6-2. In both measured and predicted temperatures, the polyester fabric had a significantly higher rate of rise than the wool fabric.

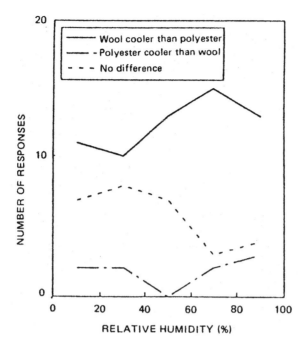

Fig. 6-1   Subjective responses to coolness to the touch [70]

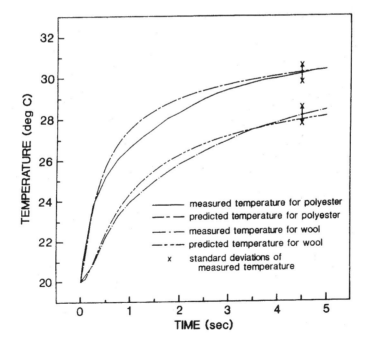

Fig. 6-2   Fabric temperature changes during fabric–skin contact [179]

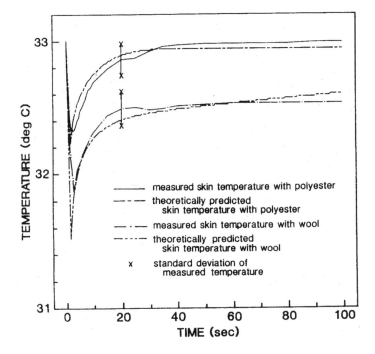

Fig. 6-3   Skin temperature changes during fabric–skin contact [179]

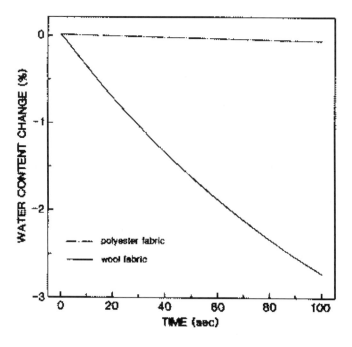

Fig. 6-4   Water content change at fabric surface during fabric–skin contact [179]

The simultaneous skin temperature changes were also compared (see Fig. 6-3). During contact, the skin temperature had a greater initial drop and smaller recovery with the wool fabric than with the polyester fabric, as shown in the measured and predicted temperature curves.

The difference in fabric and skin temperatures between wool and polyester fabric was found to be related to the moisture desorption by the fibers at the fabric surface. In Fig. 6-4, the predicted water-content change at the inner surface of the wool decreased much more quickly than that of the polyester fabric. A similar difference in water content change between wool and polyester was observed in the experimental result [39].

In these experiments, the fabrics tested were equilibrated to ambient conditions before they were momentarily brought into contact with the skin. During practical wear, garments are not in equilibrium with the surrounding environment, but are somewhere between skin and ambient conditions; the fabric comes into contact with the skin and moves away intermittently. The mathematical model was applied also to a wear condition, in which the fabric is initially 5 mm away from the skin, and is assumed to be in equilibrium with the local environment at that point; then it is brought into contact with the skin for 30 seconds, and moved 5 mm away from the skin for 30 seconds periodically. The predicted changes in the temperature and water content at the inner surface of the wool and polyester fabrics were different, as illustrated in Figures 6-5 and 6-6. These two figures indicate that coolness to the touch could be enhanced by the hygroscopicity of the fibers under practical wear conditions.

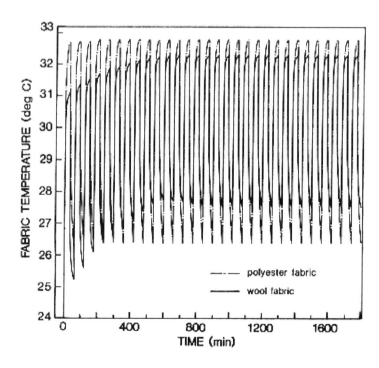

Fig. 6-5   Predicted temperature at fabric inner-surface during regular fabric–skin contact

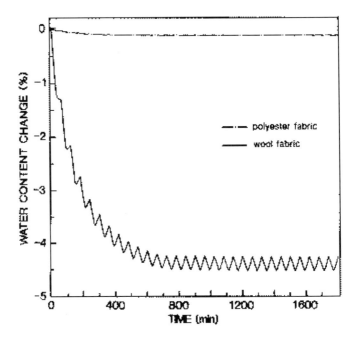

Fig. 6-6 Predicted water content at fabric inner-surface during regular fabric–skin contact

Li *et al.* (1996) [115] further investigated the mechanisms involved. The model describing the coupled heat and moisture transfer in fabrics (Section 5) was interfaced with a model of skin thermoreceptors responding to skin temperature changes (Section 3). The authors found that the change in skin temperature was related to the moisture sorption at the surface of fabric contacting the skin, as shown in Fig. 6-7. The responses from thermoreceptors depend on the skin temperature and its rate of change, which determines the subjective differentiation in coolness to the touch (Section 3).

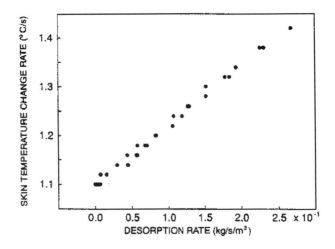

Fig. 6-7 Maximum rate of skin temperature change and moisture sorption rate of fabric [115]

The moisture desorption rate is a roughly linear function of fiber hygroscopicity and the diffusion coefficient of water vapor in fibers (see Fig. 6-8). Fiber hygroscopicity was defined as the average slope of the moisture sorption isotherm in the range of 10–80% r.h. The typical values of hygroscopicity for textile fibers vary from 0.006% for polyester to 0.205% for wool. Fig. 6-8 shows that the moisture desorption rate increases almost linearly with fiber hygroscopicity and the diffusion coefficient, suggesting that fabric coolness to the touch is a function of fiber hygroscopicity and water diffusion coefficient.

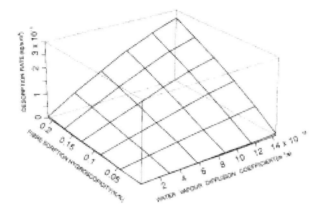

Fig. 6-8   Moisture desorption rate, fiber hygroscopicity, and water vapor diffusion coefficient [115]

The authors also studied the relationship of desorption rate with fiber diameter and fiber hygroscopicity, using the model. Fig. 6-9 shows the results under the same climatic conditions, 28°C and 70% r.h., with the water vapor diffusion coefficient fixed at 5.8 × 10^{-13} m²/s. The desorption rate had a negative, non-linear relationship with fiber diameter. Below about 15 mm, the desorption rate increased considerably with further decrease in diameter. It is clear that, by reducing fiber diameter, fabric coolness to the touch can be enhanced. However, the impact of diameter reduction is much smaller than that of hygroscopicity increase. Comparing the desorption rate of an ultra-fine and weakly hygroscopic synthetic fabric with that of normal cotton and wool fabrics, the performance of coolness to the touch still strongly favors the natural fibers.

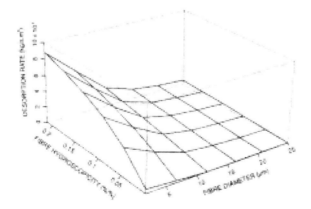

Fig. 6-9   Desorption rate with fiber diameter and hygroscopicity [115]

Sensations of coolness to the touch are most needed in warm, humid conditions. Schneider *et al.* [40] showed that fabric coolness associated with fiber hygroscopicity generally was positively related to relative humidity. Li *et al.* [115] used the model to show the influence of the temperature and relative humidity of the air surrounding the fabric on the desorption rate at the moment of maximum skin temperature change, as shown in Fig. 6-10. The desorption rate has a positive relation with relative humidity over the humidity range from 20% to 80%. However, it showed a negative relationship with ambient temperature.

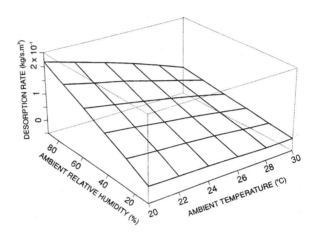

Fig. 6-10 Desorption rate with environmental temperature and humidity [115]

These results indicate the relative contributions of a number of key parameters to fabric coolness to the touch. Fiber hygroscopicity has been shown to be the most significant factor influencing the moisture desorption rate among the properties of the fibers investigated. The moisture desorption rate during contact is influenced to a lesser extent by other parameters such as the water vapor diffusion coefficient, fiber diameter, ambient relative humidity, and temperature.

These publications have shown that the moisture sorption behavior of textile fibers has significant impact on fabric coolness to the touch during fabric–skin contact. The heat and mass transfer process, which is influenced by fiber properties such as hygroscopicity and diameter, determines the temperature change on the skin surface, and hence the sensory coolness response.

On the other hand, the heat and moisture transfer process between skin and fabric is also greatly influenced by the state of the fabric–skin contact, which is determined largely by the surface features of the skin and of the fabric. Schneider and Holcombe (1991) [187] studied the fabric properties influencing coolness to the touch. They developed a three-layer model of fabric structure, as shown in Fig. 6-11. A fabric was considered as having three layers: a dense core and two outer layers consisting of a predominance of air with a small number of projecting fibers. They showed that the thickness of the outer layers had a negative influence on the rate of subjective coolness perception and temperature drop at the skin surface during the fabric–skin contact.

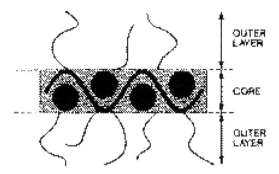

Fig. 6-11 Three-layer fabric model [187, p.491, Fig. 5]

Li and Brown [188] investigated the relationships between subjective perception of coolness to the touch and fabric properties. A range of 20 fabrics, from micropolyester lightweight fabric to wool fleece fabric, was tested. A number of statistical tools were used to identify the relative contributions of various fiber properties and fabric structural features to the perception of coolness. It was found that the subjective perception of coolness was negatively related to fabric porosity, fiber diameter, and fabric hairiness, but positively related to fiber hygroscopicity. Among these parameters, fiber hygroscopicity, fabric porosity, and fabric hairiness were the most important contributors. With matching construction, a wool fabric could be 1.5 times cooler than a polyester fabric. When the fabric porosity decreased from 0.95 to 0.65, a wool fabric could increase its coolness to the touch by 55%. If a wool fabric changed its surface hairiness from 80 count to zero count, the fabric coolness could be increased by 84%.

Kawabata and Yoneda [189–191] reported a series of theoretical studies of the heat transfer process that occurs when a fabric is brought into contact with the skin. They developed a device to measure the coolness–warmth to the touch of fabrics. The maximum heat flux during contact was proposed as a predictor of the warm–cool feel. Hes *et al.* (1990) [192] also reported an apparatus to measure the heat flow and the thermal contact properties such as heat capacity, thermal conductivity, and thermal diffusivity, which were considered to be related to the warmth of textiles. Li (1996) [193] compared the relationship between subjective perception of fabric coolness and measured criteria, such as the maximum heat flux during contact, and psycho-sensory intensity (PSI). It was found that the PSI was more closely correlated with the subjective perception than the maximum heat flux. Also, a device was developed to test the contribution of moisture desorption to the subjective coolness ratings.

6.3 Warmth

The warmth of clothing refers to three relatively independent but related aspects: the thermal insulation of clothing under steady-state and under transient conditions, and the warm sensation during fabric–skin contact. The thermal insulation of clothing under steady state conditions has been extensively studied and the results applied to practice in the last few decades; this is discussed in Section 5.1. The sensation of warmth to the touch is the opposite to the concept of coolness to the touch. The same physical mechanism applies to both warmth to the touch and coolness to the touch, the latter having been discussed in the previous section. The thermal insulation under transient conditions is related to the concept of heat of sorption, which influences the thermal insulation value of the garments and the thermal sensation to the wearer under dynamic wear situations. David (1964,1965) [151,194] investigated the thermal buffering of clothing by conducting a series of experiments. He found that 30% to 50% of the total sorption heat released was effective in reducing the heat loss through the fabric.

Olesen and Nielsen (1983) [195] measured the thermal insulation of clothing under transient humidity conditions by using a movable thermal manikin and human subjects. They observed significant differences in heat exchange between hygroscopic and non-hygroscopic fibers. Similarly, de Dear *et al.* (1987) [196] studied the thermal responses of wool clothing during humidity transients by moving a manikin dressed in a woolen garment from ambient conditions of 25°C and 20% r.h. to 25°C and 80% r.h. They found that 34% to 42% of the total heat of sorption generated was effective in influencing the sensible heat balance of the wearer. From the human subjects, they observed significant differences between wool and polyester in skin temperature changes and subjective thermal perceptions of thermal sensation, in which wool was perceived to be warmer than polyester during humidity transients.

Stuart *et al.* (1989) [152] used woolen mittens to study the perception of sorption heat. They found that the heat of sorption by hygroscopic fabrics when moved from a dry atmosphere to a moist one, is sufficient to influence wearers' perception of warmth. Mackeprang *et al.* (1990) [197] reported a similar study on sorption heat in hygroscopic clothing. By conducting wear trials, they observed that the heat changes due to moisture sorption in humidity transients were sufficiently large to be perceived readily.

6.4 Dampness

Moisture in clothing has been widely recognized as one of the most important factors contributing to discomfort sensations. Neilson and Endrusick [116] studied the influence of subjects' physical activities and clothing structure on various sensations related to temperature and moisture. They identified that the skin wetness contributed to the sensation of humidity, and that the sensation of dampness was related to the amount of sweat accumulated in clothing. The subjective sensations of skin and clothing wetness were recommended as sensitive criteria for evaluation of the thermal function of clothing.

In studying the perception approach to clothing comfort, Hollies *et al.* [15] found that a sweating sensation could be generated artificially by adding 10–20% of water into clothing. Scheurell *et al.* [198] compared the results from subjective perception of moisture with the measurements on the 'dynamic surface wetness of fabric' by using a clothing hygrometer, and found that the two were closely correlated. Hong *et al.* [199] also reported that the dynamic surface wetness was influenced by fiber types. These findings indicate that moisture in clothing contributes significantly to comfort perceptions during actual wear conditions.

Hock *et al.* (1944) [200] investigated the thermal and moisture sensations experienced by the skin. They reported that a chilling sensation was produced when damp fabrics were placed on the forearm, which was correlated with the temperature drop at an artificial 'skin' in contact with the moist fabrics. Also, the temperature drop was influenced significantly by the degree of the fabric–skin contact that was associated with fabric construction and surface hairiness. Li *et al.* [33] reported significant differences in subjective perception of fabric dampness between wool and polyester fabrics at various levels of moisture content. Plante *et al.* (1995) [34] further showed that perception of fabric dampness was a function of fabric moisture content, fiber hygroscopicity, and ambient relative humidity.

These findings indicated that the process of perception of dampness involves complex mechanisms which may be related to the dynamic heat and moisture processes in fabrics.

Therefore, a series of studies was carried out by Li *et al.* (1993,1995) [33,121] to investigate the physical mechanisms in dampness perception. The mathematical model described in Section 5 was used to study the dynamic heat and moisture transfer processes in detail during the course of the perception. A physiological trial was also carried out to obtain the skin temperature changes during contact between moist fabrics and the skin. The subjective rating scores on fabric dampness perception were found to be a function of skin temperature drop (see Fig. 3-17). The changes in skin temperature and fabric temperature during the contact are shown in Fig. 6-12; readings were obtained from both the theoretical predictions and experimental observations.

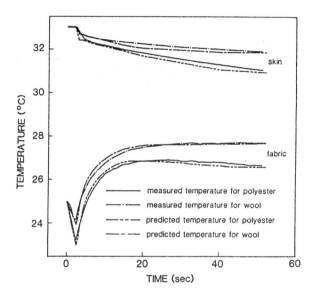

Fig. 6-12 Temperature changes during the fabric–skin contact [33]

The skin and fabric temperature changes were shown to be related to the changes in moisture content of the fabrics. As illustrated in Fig. 6-13, the rates of moisture change were significantly different between wool and polyester fabrics. This is related to the mechanisms of moisture changes between the fibers and their surrounding air, as outlined in Fig. 5-2, and described by Equations (5-9) and (5-12). At 4% of excess water above equilibrium regain, polyester fabric had a total moisture content of about 5%, which was well above the saturation regain of the fiber. The moisture exchange process was a surface evaporation following Equation (5-9).

At the same excess water content, the total water content of wool fabric in the test condition was about 11%, which was well below its saturation regain. Thus, the moisture exchange process was largely a desorption process following Equation (5-12), in which moisture desorbs from within the fibers and then diffuses through the fabric to the surroundings. This means that the actual heat and moisture transport processes in the fibers and fabrics determine the heat exchange between the skin and fabrics, which in turn determines the magnitude of the physical stimulus and influences the subjective perception of dampness.

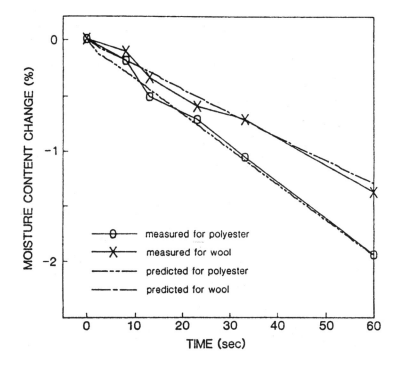

Fig. 6-13 Changes in fabric moisture content for wool and polyester during the process of dampness perception [33]

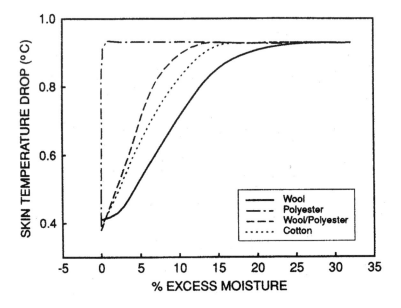

Fig. 6-14 Fiber hygroscopicity, excess moisture, and the skin temperature drop

As fiber hygroscopicity is a critical factor determining the coupled heat and moisture transfer behavior in fabric, it has a significant impact on the skin temperature drop during the contact. By using the model, the authors showed that the skin temperature drop during a fabric–skin contact increases progressively with increasing level of excess moisture, and reaches a maximum magnitude when the constituting fiber reaches its saturation moisture content, as shown in Fig. 6-14. Comparing fabrics with different degree of hygroscopicity, the skin temperature drop increases more slowly with the level of excess moisture as the degree of fiber hygroscopicity increases.

Ambient conditions such as temperature and relative humidity influence the skin temperature drop significantly. As shown in Fig. 6-15, the skin temperature drop decreases as ambient temperature increases, because of the decrease in temperature difference prior to the contact. However, ambient temperature has negligible influence on the differences of the skin temperature drop among different types of fibers, because ambient temperature mainly influences the dry heat transfer process, not the moisture exchange process.

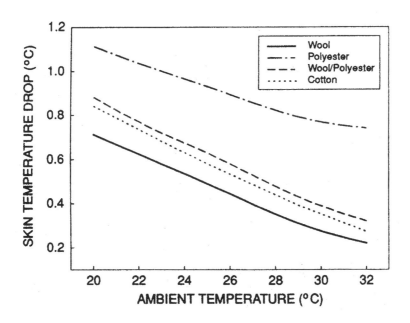

Fig. 6-15 Influence of ambient temperature on the skin temperature drop

Ambient relative humidity, on the other hand, shows significant impact on both the skin temperature drop of all fibers and the differences between the fibers (see Fig. 6-16). When ambient relative humidity increases, the difference in moisture concentration between the fabric and the environment decreases, resulting in a smaller temperature gradient between the skin and fabric, hence a smaller skin temperature drop during the skin–fabric contact. The differences between fibers are much greater when ambient relative humidity is low. When the relative humidity approaches saturation, the differences between fibers becomes negligible.

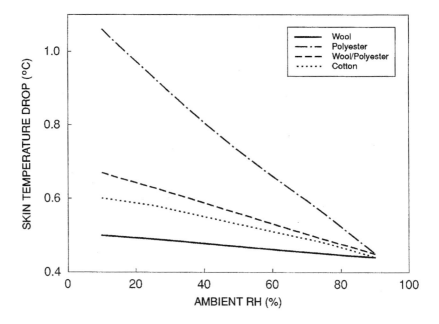

Fig. 6-16 Influence of ambient relative humidity on the skin temperature drop [121]

6.5 Clamminess and Moisture Buffering During Exercise

In 1939, Cassie *et al.* [201] first postulated the concept of moisture buffering of clothing. Cassie, in 1962 [202], restated the hypothesis that the sorption of moisture by hygroscopic clothing would release heat, which would have significant impact on the heat balance and thermal perceptions of the wearer experiencing a sudden change from a warm, dry atmosphere to a cold, damp one.

Hygroscopic fibers have the ability to absorb a considerable amount of moisture from the surrounding atmosphere. With humidity transients, hygroscopic fibers can absorb or desorb moisture from, or to, the adjacent air, which can delay the moisture change in the clothing microclimate. Theoretically, this effect often acts as a buffer against sudden humidity changes in favor of the wearer. However, the effectiveness of such a buffering effect varies widely as reported in the literature. Rodwell *et al.* (1965) [203] investigated the physiological impact of sorption heat in clothing. They did not find significant impact of sorption heat in the simulated damp, cold winter conditions. Umbach (1980) [204] studied the buffering effects of synthetic and cotton underwear. Wear trials were conducted with subjects walking at 7 km/h at ambient conditions of 27°C and 50% r.h. No significant differences were found in moisture buffering behavior between underwear made from the two fibers. Therefore, they concluded that there was no evidence indicating that moisture sorption of fibers has a decisive role in the comfort of clothing worn next to the skin.

Similar results were obtained from various wear trials conducted by Andreen *et al.* (1953) [205], Vokac *et al.* (1976) [206], Holmer (1985) [207], and Hatch *et al.* (1990) [208]. Various physiological parameters were measured in those trials, such as sweating rate, stratum corneum water content, water evaporation from the skin, energy output, microclimate temperature, and humidity. No significant differences were found between fabrics made from highly hygroscopic fibers such as wool, and those made from weakly hygroscopic fibers such as polyester and polypropylene.

On the other hand, a number of research publications have reported significant differences in moisture buffering between hygroscopic and non-hygroscopic fibers. Spencer-Smith (1978) [209] illustrated the buffering effect of hygroscopic clothing by using a simple, approximate, graphic explanation. Using a simulated sweating device, Scheurell *et al.* (1985) [198] examined the response of fabrics to humidity and temperature gradients. The authors reported that a fabric property, identified as the dynamic surface wetness, correlated with the comfort perception during wear.

In a study of the influence of cotton, polyester, and blended fabrics on dynamic surface wetness and moisture transfer through textiles, Hong *et al.* [199] found that the moisture build-up at the inner fabric surface facing the sweating skin was the slowest with all-cotton fabrics, followed by cotton–polyester blends, and all-polyester fabrics. The ranking order corresponded with the level of fiber hygroscopicity. Wehner *et al.* (1988) [210] developed an apparatus to simultaneously measure the moisture sorption by a fabric and the moisture flux through a fabric during humidity transients. By comparing fabrics with different level of hygroscopicity, the authors found that the duration of the transient behavior depended strongly on the moisture sorption capacity of the fabric. The moisture flux across an inert porous barrier can reach a steady state within seconds, while non-steady condition may last for more than an hour when a wool fabric is exposed to a humidity gradient. During the transient period, the total amount of moisture removed from a high-humidity environment is greater with a highly hygroscopic fabric such as cotton than with a weakly hygroscopic fabric such as polyester.

In comparing wool and polyamide clothing, Behmann (1971) [211] conducted a wear trial using one subject repeatedly walking on a treadmill at 2.2 km/hr under conditions of 30°C, 30% and 60% r.h. The author reported that the evaporation limit was reached later when the subject wore a wool garment, and the perception of sweating and clinging sensations were delayed. Tokura *et al.* (1982) [212] conducted a wear trial in a hot environment (33°C and 60% r.h) to study the effect of moisture absorbency of fibers on the sweating rates of secondary subjects. It was found that subjects wearing polyester lost greater amounts of sweat, and the humidity in the microclimate was significantly higher when wearing cotton.

Li *et al.* (1992) [32] compared wool with polyester by carrying out wear trials at 28°C and 30% r.h. Subjects were walking on a treadmill at 5.6 km/hr for up to 40 minutes. No difference between wearing wool and polyester was found in sweat loss from the upper body during exercise. Wool garments, however, took up significantly more sweat than polyester garments. Correspondingly, subjects reported feeling less clammy, and warmer, when wearing wool than when wearing polyester during a period of between 10 to 30 minutes of exercise, as shown in Figures 6-17 and 6-18. The coupled heat and moisture transfer model (Section 5) was applied to these wear situations by specifying boundary conditions that described the situations. The model predicted that the relative humidity of the microclimate would be lower and skin temperature would be higher during the 10 to 30 minutes of exercise.

The authors further applied the model to analyze the buffering effect of hygroscopic clothing under previously reported wear trials, and used it to explain the contradictory results reported in the literature. The model indicated that the moisture buffering by hygroscopic fibers could be effective during a certain period after exercise. The length of this buffering period and the magnitude of delay of humidity rise depend on the ability of the fabric to remove moisture relative to the speed of moisture build-up in the clothing microclimate, which is related to ambient conditions, clothing material and style, and exercise intensity of the subjects. Therefore, the apparent contradiction on clothing buffering effect reported in the literature can be largely attributed to the differences in the climatic and exercise conditions used.

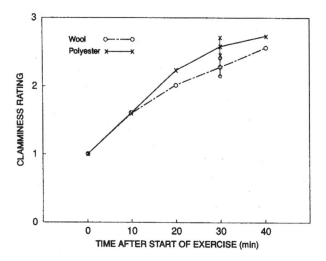

Fig. 6-17 Perception of clamminess during exercise [32]

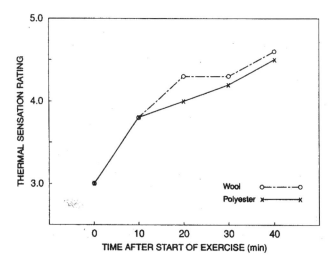

Fig. 6-18 Perception of warmth during exercise [32]

In a study of the influence of fiber hygroscopicity on the thermoregulatory responses during exercise, Li and Holcombe (1993) [57] compared wool with polyester by conducting a wear trial at 20°C and 35% r.h. Subjects were exercised by peddling on a cycle ergometer at a load adjusted to achieve a heart rate approximately 70% of the age-weighted maximum for each individual. It was found that the relative humidity measured in the microclimate of their clothing rose more quickly when wearing polyester than when wearing wool, as shown in Fig. 6-19. The difference between the two was significant in the period between 10 to 20 minutes after exercise.

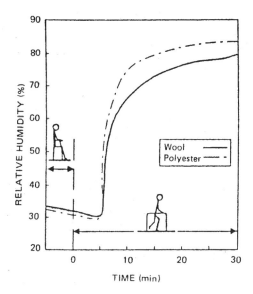

Fig. 6-19 Relative humidity measured in the microclimate of clothing

From the corresponding temperatures measured at the skin and fabric surface, the authors observed significant differences between wearing wool and polyester. After sweating, the temperature of the garment, and the corresponding skin temperature, rose significantly more when wearing wool than when wearing polyester, as shown in Fig. 6-20. The difference is statistically significant for the first 20 minutes after sweating starts, following which the difference between the two gradually diminishes. On the other hand, the mean core temperature showed no significant difference when wearing wool and polyester, even though the core temperature seemed to rise more slowly when wearing wool.

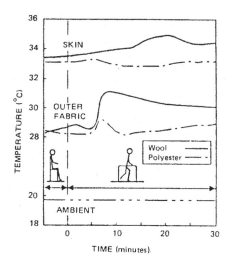

Fig. 6-20 Temperature changes at the skin and fabric surfaces during exercise

The authors calculated the dry and latent heat fluxes at the outer surface of the garment according to the measured temperature and moisture gradients. As illustrated in Fig. 6-21, no difference between dry and latent flux was observed before sweating. After sweating, the dry heat flux at the outer surface of the garment was significantly higher when wearing wool than when wearing polyester. No significant difference in latent heat flux was found between wool and polyester. The total heat flux at the outer surface of a garment after sweating was about 13% greater with wool than with polyester until the end of exercise period; this was significant at 0.01 level.

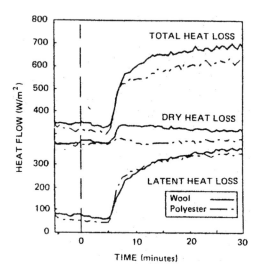

Fig. 6-21 Heat fluxes at the outer clothing surface during exercise

These results indicate that fiber hygroscopicity has a significant impact on the thermal response of the body and the heat balance of the body–clothing system during the transient period of exercise. When sweating starts, highly hygroscopic fibers absorb considerable amounts of sweat and their temperature rises due to the heat of sorption released. The elevated fabric temperature interacts with the body, stimulating higher skin temperature and raising sweat rate.

Some of the sweat is further absorbed by the fabric, adding to the release of sorption heat and increasing the dry heat loss at the outer surface of the garment. Hence, the body is able to shed more heat during exercise. The sorption of moisture and the released sorption heat by weakly hygroscopic fibers such as polyester are very low. Most of the sweat in the garment was present as liquid and it had a smaller influence on the dry heat loss at the outer surface of the garments. Therefore, the role of clothing made from weakly hygroscopic fibers is more passive and its enhancement of heat loss during exercise is smaller.

6.6 Environmental Buffering

The buffering effect of hygroscopic clothing has significant impact on the thermal balance and comfort of the wearer during the humidity transients due to environmental changes. Often, a wearer experiences various sudden and large changes in the external thermal environment. For instance, a wearer may be exposed to differences in temperature and humidity greater than 10°C and 30% r.h. when walking from an air-conditioned indoor environment to a hot and humid summer outdoor environment. The difference in temperature between an air-conditioned indoor environment and an outdoor winter environment in the cold regions can be greater than 20°C. Clothing is an extremely important barrier to protect the body against such sudden environmental changes.

An investigation of the buffering effect of clothing against rain was reported by Li (1977) [158]. The author analyzed the physical process of a rain droplet contacting the surface of a garment and identified that there are significant differences in temperature and humidity between the dry clothing surface and rain droplets. The differences are a function of ambient temperature and ambient r.h. The lower the ambient r.h., the greater the temperature difference. Due to these differences, when rain droplets impact on a garment the temperature at the clothing outer surface can decrease and relative humidity will rise to 100%.

By specifying such dynamic changes in the boundary condition at a clothing surface, the magnitude of the buffering effect for a single-layer garment made from wool and polyester was estimated by using a model simulating the heat and moisture transport processes in clothing and their interaction with the thermoregulatory system (Sections 4 and 5). The temperature changes at the skin surface when a person wears wool or polyester garments in an ambient temperature of 25°C and an r.h. varying from 30% to 90% and is suddenly caught in the rain, were predicted and are shown in Fig. 6-22. The skin temperature changes when wearing wool were predicted to be smaller than those when wearing polyester. As ambient relative humidity increases, the temperature changes become smaller when wearing both fibers. Also, the differences in skin temperature changes between wool and polyester decrease with increasing ambient relative humidity.

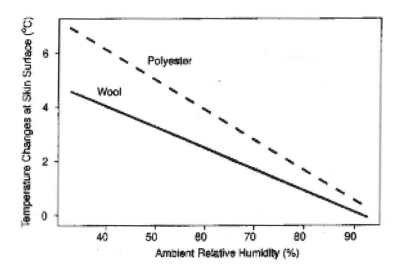

Fig. 6-22 Prediction of skin temperature changes during rain

These predictions suggest that there are differences between wool and polyester fibers in buffering against rain-chill, which represent the extremes of hygroscopicity. For other textile fibers such as cotton and acrylic, whose hygroscopicities fall between these extremes, the effectiveness of buffering is expected to be related to the extent of their hygroscopicity. The more hygroscopic the fiber, the stronger the buffering effect.

In order to test these predictions, a series of physiological wear trials was carried out using woolen and acrylic jumpers. The temperature profile from skin surface to inner clothing surface to outer clothing surface was measured, as shown in Fig. 6-23. At the onset of rain, the temperature at the outer surface of the jumpers, where the raindrops hit directly, fell by approximately 3–4°C within one minute. With wool, the temperature drop tended to be smaller than with acrylic fiber. At the inner surface, the difference between wool and acrylic fiber was more obvious. With the latter, a rapid temperature drop (about 1.2°C within a minute) was observed, while with wool the temperature decreased gradually over 10 minutes. These differences were extended to the skin surface level. When a subject was wearing acrylic, the skin temperature decreased more than when he was wearing wool (significant at the 98% level).

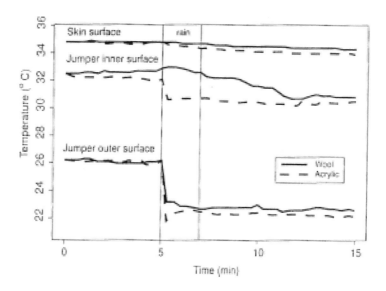

Fig. 6-23 Temperature profile in clothing during rain

The corresponding relative humidity at the skin surface and at the inner surface of both jumpers was also measured, as illustrated in Fig. 6-24. At both surfaces, the r.h. increased rapidly during the first minute of rain, then continued to rise slowly. The differences in the humidity increase between wool and acrylic were significant at the 99% level after rain. When wearing acrylic, the increase in humidity was greater and faster than when wearing wool. These experimental results confirmed the predictions that highly hygroscopic fibers such as wool could buffer the body against sudden changes in temperature and humidity more effectively.

Further, a series of psychological wear trials was conducted to test whether the buffering effect was perceivable or not. Subjective ratings on warmth, dampness, and comfort over 20 subjects were obtained. Significant differences in the subjective ratings were found between woolen and acrylic jumpers. Perception of warmth deceased and perception of dampness increased considerably as rain started, and slowly recovered after rain ceased, as shown in Fig. 6-25. The drop in perception of warmth (Fig. 6-25(*a*)) and increase in dampness sensation (Fig. 6-25(*b*)) were considerably greater for acrylic jumpers than for woolen jumpers, indicating that the impact of rain on thermal, moisture and comfort perception was perceivable and that woollen jumpers had a stronger buffering effect than acrylic jumpers.

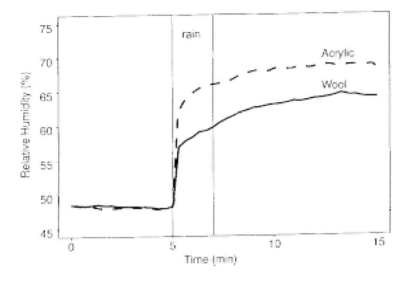

Fig. 6-24 Relative humidity in clothing microclimate during rain

Simultaneously, the overall perception of comfort decreased quickly during the raining period, and continued to decrease afterwards (Fig. 6-25(*c*)). Significant differences were observed between wool and acrylic. When wearing acrylic jumpers, the overall comfort ratings declined more rapidly during the rain period than when wearing woolen jumpers, indicating that the buffering effect of wool against temperature and moisture change had an impact on perception of overall comfort of the subjects.

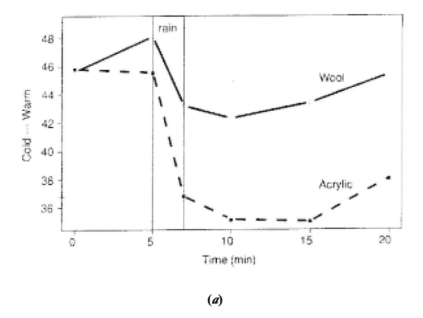

(*a*)

Fig. 6-25 Perception of warmth, dampness, and comfort in rain. (*a*)Warmth; (*b*) Dampness; (*c*) Comfort

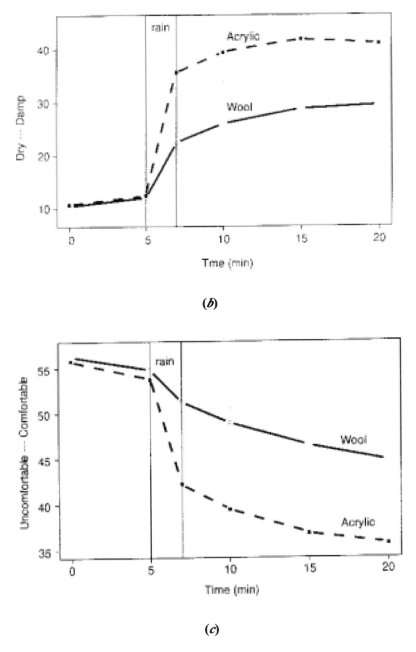

Fig. 6-25 Perception of warmth, dampness, and comfort in rain. (*a*)Warmth; (*b*) Dampness; (*c*) Comfort

7. FABRIC MECHANICAL PROPERTIES AND TACTILE-PRESSURE SENSATIONS

7.1 Fabric Prickliness

Unlike the area of clothing thermal and moisture comfort, the theoretical framework on the physical mechanisms of clothing tactile and pressure comfort have not been fully developed. However, a considerable volume of research outcomes has been reported on the relationships between fiber-to-fabric properties and various tactile and pressure sensations.

Fabric-evoked prickle has been identified as one of the most irritating discomfort sensations for clothing wear next-to-skin. The neurophysiological basis of fabric prickle perception has been well established; as discussed in Section 3.3.3. Garnsworthy *et al.* (1988) [84] identified a special type of pain nerve responsible for prickle sensation, which is triggered by a threshold of force of about 0.75 mN. Individual protruding fiber ends from a fabric surface are responsible for triggering the pain nerve endings when contacting the skin. Summation of responses from a group of pain nerves seems necessary for the perception of prickle sensations. In subjective tests, Garnsworthy *et al.* [86] observed that most of the subjects could not perceived prickle from fabrics containing a density of high load bearing fiber ends less than 3 per 10 cm². The intensity of fabric prickle perception is a function of the density of high load bearing fiber ends at the fabric surface and the area of contact between the fabric and skin. This indicates that both fiber mechanical properties and fabric surface features are important factors determining fabric-evoked sensations.

Matsudaira *et al.* (1990) [213] compared three techniques for objective measurement of fabric prickle: low-pressure compression testing, laser-counting of protruding fibers, and a modified audio-pickup method. A KES-FB compression tester was modified to measure the relationship between pressure and fabric thickness at the initial compression stage, in which the protruding fibers are bent and compressed. A laser hairiness meter developed at WRONZ was used to count the fibers protruding from the fabric surface. The sensitivity of the instrument was found inadequate for the detection of all fabric surface hairs, with a bias in favor of the coarser and stiffer hairs. An audio pick-up head was therefore modified to detect the protruding fibers from a fabric surface, as shown in Fig. 7-1. They found that the modified audio pick-up technique was the most effective and the measured mean force per contact correlated well with the subjective perception of fabric prickle.

During testing, the fabric surface was traversed under a stationary audio stylus, from which signals were obtained from the contact between the stylus and a protruding fiber. In developing a calibration of the stylus signal, the authors used two classical models — a loaded cantilever and an Euler column — to calculate the pointing force and the critical buckling load, as shown in Fig. 7-2.

The critical buckling load, P_E, of the protruding fiber ends has been identified as the stimuli responsible for triggering the pain receptors, and can be expressed as:

$$P_E = \pi^2 \left(EI/4l^2 \right)$$

where E is the Young's modulus of the fiber;
I is the moment of inertia ($I = \pi d^4/64$ in the case of a circular rod); and
l is the length of the protruding fiber ends.

This equation clearly suggests that fiber Young's modulus, fiber diameter, and fiber length are the key factors determining fabric prickliness. Veitch and Naylor (1992) [214] studied the mechanics of fiber buckling in the fabric prickling process and concluded that the short fiber

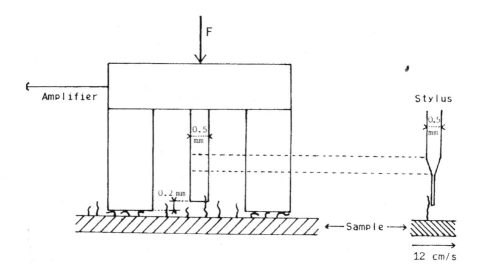

Fig. 7-1 Audio pick-up device for measurement of fabric prickle [213]

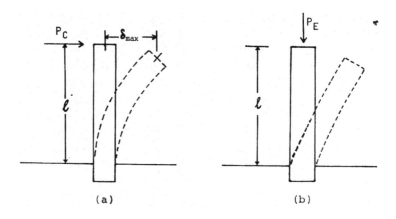

Fig. 7-2 Models of (*a*) a loaded cantilever, and (*b*) an Euler column [213]

protruding ends obey Euler's simple buckling theory. By assuming a nominal length of 2 mm, Garnsworthy *et al.* [86] and Naylor [215] calculated that a fiber with a diameter of 30 mm gives a buckling load of 0.75 mN. Using a forearm test method, the authors evaluated the prickliness of a set of wool knitted fabric samples covering a range of fiber diameters from 19 mm to 30 mm. By analyzing the relationships between subjective prickle perception and fabric features such as fiber diameters, treatment, and finishing, they derived a multiple linear regression equation:

$$MPE = -3.65 + 2.83 \text{ (diameter)} - 0.60 \text{ (treatment)} - 0.25 \text{ (finish)},$$

where *MPE* was the mean prickle estimate. The treatment was coded 0 for untreated fabric and 1 for Kroy/silicone treated fabric. The finish was coded as 0 for steam relaxation and 1 for an aqueous scour. This equation suggested that subjective perception increased with fiber diameter and decreased with reduction of fiber-fiber friction by antiprickle treatment and finishing processes.

Matsudaira *et al*. [216] investigated the effects of finishing on fabric prickle. They found that, for the fabric surfaces containing 35 mm wool fibers, the decreases and increases in prickle due to successive finishing processes were appreciable. By blending acrylic fibers with different diameters, Naylor (1992) [215] studied the effect on prickle of the composition and shape of the diameter distribution of the coarse-edge fibers. The author observed that the prickle of single jersey knitted fabric correlated with the percentage of fibers with diameters greater than a threshold value, which was close to 30 mm. Also, wool and acrylic fabrics with similar diameter distributions were shown to have the same level of prickle. In a further study of fabric-evoked prickle, Naylor *et al*. (1997) [217] used a range of worsted spun single jersey knitted fabrics made from wool with different fiber diameter distributions. The authors reported that the prickle sensation of these fabrics could be predicted from the density of coarse fiber ends per unit area of fabric. The percentage of fiber ends with diameter greater than 32 mm was identified as the key factor. Also, the variations in diameter distribution in individual fleeces, especially the percentage of coarse fiber ends, influence fabric prickle significantly.

Kennis (1992) [218] reported a study on the influence of some parameters on prickle evoked by woven fabrics. A series of weft sateen fabrics was made from two types of wool: 18.2 mm and 23 mm. Fabrics were manufactured into high and low cover factors with weights around 250 $g\,m^{-2}$. All the fabrics were scoured, centrifuged, hung to dry, cropped, and blown for 2 minutes. The three most influential factors were identified as fiber diameter, fabric cover factor, and finishing. The interactions between the factors were not significant. In an investigation into the influence of fabric mechanical properties on the discomfort sensations in wear trials, Li and Keighley (1988) [219] found that fabric prickle sensation was positively correlated with fiber diameter, fabric thickness at low loading, and fabric surface roughness.

7.2 Fabric Itchiness

Similar to fabric prickle, itch is also found to result from activation of some superficial pain receptors [87]. A prickling fabric usually has a quality of itch sensation [86]. In a number of psychological wear trials, it was found that the perception of itchiness in clothing was highly correlated with the perception of prickliness. Both sensations were classified in the tactile sensory factor [65,41]. Therefore, it could be expected that the factors influencing fabric prickle would effect fabric itchiness as well. Comparing the subjective ratings of itchiness obtained from wear trials with the mechanical properties measured objectively, Li (1988) [220] observed that perception of itchiness correlated with fiber diameter, fabric thickness at low and high pressures, and fabric surface roughness.

7.3 Fabric Stiffness

Elder *et al*. (1984) [221] studied the stiffness of woven and non-woven fabrics by using both subjective assessment and objective measurements. Magnitude estimation was used to obtain the subjective responses. Objective measurements of fabric stiffness were carried out by using a Shirley Cantilever, a Cusick Drapemeter, and a Shirley Cyclic Bending Tester. The authors observed that the flexural rigidity obtained by bending-hysteresis measurements using a Shirley Cyclic Bending Tester correlated well with the subjective estimation of fabric stiffness. Applying Steven's law to the data, the authors found a logarithmic linear relationship between the subjective stiffness estimation and the flexural rigidity. Further, Elder [60] reported a study to verify their methodology and conclusion by using another set of woven fabrics and knitted fabrics. The author found that the agreement among three objective measurements — bending length, flexural rigidity, and drape coefficient — was good, and that these measurements were highly correlated with the subjective ratings.

One-year later, Elder *et al.* (1985) [134] described a psychological scale for fabric stiffness. In this paper, the authors selected drape coefficient, measured through the Cusick Drapemeter, as the objective measure of stiffness, instead of the flexural rigidity measured by the Cantilever. The main reason for using the drape coefficient was that the criterion could give an integrated measure, similar to that of a human being. The Cantilever was rejected because the measurements were directional and had greater variability. Again, a good logarithmic linear relationship between the subjective stiffness estimation and the drape coefficient was obtained.

Fabric stiffness has been recognized as one of the primary hand expressions used by Kawabata and Niwa in their fabric hand evaluation system named KOSHI in Japanese [21]. Using a stepwise regression method, Hu *et al.* [135] found that fabric stiffness (called HV1) was related to a number of objective parameters measured by KES-F instruments, namely WC (energy in compression of fabric under 5 kPa), B (bending rigidity), MIU (coefficient of steel–fabric friction), MMD (mean deviation of MIU), and LC (linearity of compression thickness curve).

Bishop [54] made a comprehensive summary of the objectively measurable physical properties associated with stiffness commonly used in the literature, including bending stiffness, thickness, areal density, shear stiffness/hysteresis, and compressibility. Comparing the subjective sensory responses from wear trials and the mechanical properties measured objectively, Li (1988) [220] found that subjective ratings of garment stiffness are related to three types of mechanical properties: (i) fiber diameter and tensile breaking load; (ii) fabric compression properties such as thickness at low and high pressure, the energy of the compression–thickness curve, the slope of the compression–thickness curve, and the resilience of the compression–thickness curve; and (iii) fabric frictional properties such as the mean friction coefficient and the mean deviation of friction coefficient.

7.4 Fabric Softness

Fabric softness is one of the most frequently used terms in describing clothing comfort performance by consumers. Fabric softness has multiple meanings that can be related to compression and/or to smoothness and flexibility of fabrics, depending on the fabrics being handled and end-uses. Peirce (1930) [222] considered softness as the opposite of stiffness that could be measured by bending length. Later, Howorth (1964) [19] took softness as the opposite of firmness or hardness measured by thickness tests.

Elder *et al.* (1984) [42] accepted the definition of 'ease of yielding to pressure' and conducted subjective finger-pressure assessments of fabric softness using a magnitude estimation method. Meanwhile, measurements of compression were carried out as the objective measure of fabric softness using an Instron Tensile Tester fitted with a compression load cell. The authors found that the relationship between the subjective assessment of fabric softness and the objective measurement of compression followed Stevens' law of logarithmic linear relation. The perception of softness was highly correlated with fabric compression, which was defined as the decrease in intrinsic thickness with an appropriate increase in pressure. The intrinsic thickness is the thickness of the space occupied by a fabric subjected to barely perceptible pressure. Further, the subjective softness assessment correlated with fabric thickness for woven and nonwoven fabrics, and correlated with fabric density and specific volume for woven fabrics but not for nonwoven fabrics.

In Kawabata's hand evaluation system, fabric softness was not considered as one of the primary hand values. Corresponding to 'NUMERI', softness was defined as a mixed feeling coming from a combination of smooth, supple, and soft feeling. The typical fabric for this definition is a fabric woven from cashmere fibers. In the dimension of 'FUKURAMI', softness is related to the feeling from a combination of bulky, rich, and well-formed impressions. A springy property in compression and thickness, together with a warm feeling, is also associated with softness. Corresponding to 'SOFUTOSA', softness is a feeling coming from higher 'NUMERI' and 'FUKURAMI' and weaker 'KOSHI' (stiffness).

FUKURAMI, which was interpreted as fullness and softness in English, is one of the primary-hands defined by Kawabata *et al.* Hu *et al.* [135] found that FUKURAMI is closely related to fabric thickness at low pressure (T_0), coefficient of steel-fabric friction (MIU), geometric roughness (SMD), and energy in compressing the fabric (WC).

Bishop reviewed literature and summarized the physical properties associated with softness as bending, compression and tensile properties, shear stiffness and hysteresis, areal density, and friction [54].

Li (1998) [220] observed that subjective perception of garment softness during wear correlated with fabric compression properties (thickness at low and high pressures, resilience, and energy of the compression–thickness curve), fabric tensile properties (the maximum elongation, and linearity of the load–elongation curve), fiber diameter, and breaking load. These reflect the three aspects of fabric softness identified by previous researchers: compression, flexibility, and smoothness.

7.5 Fabric Smoothness, Roughness, and Scratchiness

As a fabric is moved across the skin, displacement of skin is increased and the perception of fabric roughness or smoothness is evoked. The friction and mechanical interaction between fabric and skin during contact are the key factors determining the perception of roughness, smoothness, and scratchiness. It has been identified that roughness and scratchiness are important tactile sensations determining the comfort performance of next-to-skin wear. The friction between skin and fabric is smaller with a fabric having a smooth surface than with a fabric having a rougher surface. Moisture at the skin surface can alter the intensity of fabric roughness perceived. As moisture content increases, the friction and displacement of skin increases, which triggers more touch receptors. Therefore, a fabric that is perceived to be comfortable under low humidity conditions may be perceived to be uncomfortable under higher humidity conditions or sweating conditions.

Behmann (1990) [105] reported a study on the perception of roughness and textile construction parameters. The roughness was defined as the irregularities in the surface that can be described geometrically by the size of the roughness elements or mechanically by the friction coefficients. A roughness model was reproduced as shown in Fig. 7-3. A series of subjective perception trials were conducted using single nylon filament woven and knitted fabrics. The author concluded that the perception of roughness was determined by the roughness spacing.

Further experiments were carried out to study the influence of practical textile parameters on roughness perception, by using woven and knitted fabrics made from nylon yarns of different diameters. As shown in Fig. 7-4, the roughness perception decreased with yarn diameter logarithmically, and, at the same yarn diameter, the knitted fabric was perceived as 'rougher'.

Comparing the subjective sensory responses from wear trials with objective measured mechanical properties, Li (1988) [220] found that the perception of roughness correlated with fabric surface roughness (maximum force, mean surface roughness coefficient, and deviation of surface roughness coefficient), compression properties (fabric thickness at high and low pressures, and energy of the compression-thickness curve), fiber diameter and fiber tensile properties (breaking load and breaking elongation), and fabric tensile properties (maximum tensile elongation, elongation recovery load).

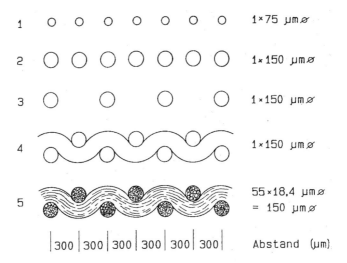

Fig. 7-3 Roughness model of fabrics [105]

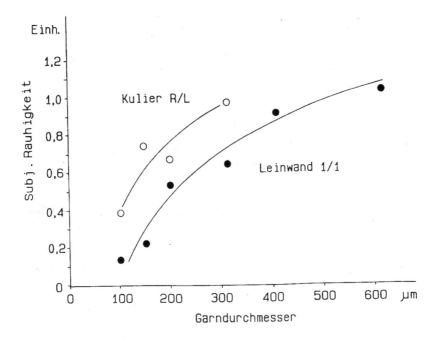

Fig. 7-4 Subjective perception of roughness as a function of the yarn diameter [105]

Similarly, subjective perception of scratchiness is related to fabric tensile properties (maximum tensile elongation, energy of the tensile load–elongation curve, and the slope of the tensile load–elongation curve), fabric surface roughness (maximum roughness force, mean surface roughness coefficient, and deviation of surface roughness coefficient), fabric compression properties (thickness at low and high pressure, linearity of the compression curve, energy of the compression-thickness curve, and slope of the compression–thickness curve).

Ito (1995) [223] studied the wear performance of girdles, and identified pressure and hand (smoothness and softness) as the important comfort properties. Changes in pressure, related to standing and moving during wear, correlated with the biaxial extension and stress relaxation properties of the fabrics used to make the girdles.

In summarizing the findings from the literature, Bishop (1996) [54] showed that perception of fabric roughness (smoothness) is associated with a number of physical properties objectively measured, such as roughness, friction, prickle, shear and bending stiffness, thickness, and areal density. In Kawabata's KES system, fabric smoothness is an important primary hand, called 'NUMERI', which is defined as a mixed feeling coming from a combination of smooth, supple, and soft sensations. The typical fabric was identified as a woven fabric made from cashmere. Hu *et al.* (1993) [135] reported that fabric smoothness is related to fabric thickness at low pressure (T_0), geometric roughness (SMD), bending rigidity (B), linearity of the compression thickness curve (LC), energy in extending the fabric to 5 N/cm (WT), fabric mass per unit area (W), hysteresis of bending moment (2HB), and energy in compressing the fabric under 5 kPa (WC).

Ajayi (1992) [224] reported a study of fabric smoothness using a friction measurement on a flat horizontal platform attached to an Instron load cell. He found that the number of peaks in the stick–slip traces per 5 cm sled traverse was linearly related to the courses/5 cm in knitted fabrics and threads/5 cm in woven fabrics. Ramgulam *et al.* (1993) [225] compared the method of measuring fabric surface roughness using a laser sensor with a conventional contact method (KES tester). Relatively good correlation between the two methods was obtained with $r = 0.801$.

Wilson and Laing (1995) [106] reported a study of the effect of wool fiber parameters on the tactile characteristics of woven fabrics. The fabrics were made from various wool fibers but constructed from a standardized yarn structure, dye, and finishing treatment. Twenty-four female subjects and 20 male subjects were used. Significant differences were found in the rankings of roughness and prickliness among the fabrics. Fabrics with fiber diameters of less than 23 microns and fiber bulk greater than 32 cm³/g were perceived to be smoother and less prickly than average. With fiber diameters of greater than 34 microns and fiber bulk less than 21 cm³/g, fabrics were perceived to be rougher and pricklier.

7.6 Garment Fit and Pressure Comfort

Consumers have an inherent desire to dress comfortably in attractive garments, which requires a reduction of garment restraint imposed on the body and an increase in the ability of fabric to 'give'. This means that the garment needs to be cut neatly and to be able to maintain a reserve of comfort for the wearer's dynamic movements. Kirk and Ibrahim (1966) [226] reported a study of the relationship between fabric extensibility and anthropometric requirements of garments. In analyzing the anthropometric kinematics, the authors identified that there are three essential components to meet the skin strain requirements: garment fit, garment slip, and fabric stretch. Garment fit provides the space allowance for skin strain, which is affected by the ratio of garment size to body size and the nature of garment design. Garment slip, which is determined mainly by the coefficient of friction between skin and fabric and between different layers of garments, is another mechanism for a garment to accommodate skin strain. Fabric stretch, an important factor for pressure comfort, is dependent largely on fabric elastic characteristics and elastic recovery properties. Whether a garment slips or stretches is dependent on the balance of the tensile forces in the fabric and the frictional forces between skin and fabric. If a fabric has a low resistance to stretch and high friction against the skin or fabric, it tends to stretch rather than slip. The opposite is true if the fabric has low friction and high tensile resistance. If a fabric has both high friction resistance and stretch resistance, high clothing pressure is likely to be exerted on the body, which will result in discomfort sensations.

The authors identified the critical strain areas of the body as the knee, the seat, the back, and the elbows. Maximum local skin strain was measured by drawing a series of lines on the skin at regular intervals and measuring changes in skin dimensions that took place with critical body movement. Table 7-1 summarizes the measurements of skin strain at the identified areas.

Table 7-1
Skin Strains at Various Critical Body Areas [226]

Body element	Body movement	Local skin strain (%)			
		Horizontal		Vertical	
		Men	Women	Men	Women
Knee	Stand → sit	21	19	41	43
Knee	Stand → deep bend	29	28	49	52
Elbow	Straight → full bend	24	25	50	51
Seat	Stand → sit: overall (hip to hip)	20	15	27	27
Seat	Stand → sit: local crotch	42	35		
	local buttocks			39	40
Seat	Stand → bend: overall	21	17	27	27
Seat	Stand → bend: local crotch	41	37		
	local buttocks			45	45
		Local skin strain (%)			
Back	Straight → forward raised arm	33	31		
	→ elbows on table	28	28		
	→ elbow bending	16	16		
	→ shoe tying	47	47		

The results showed that the skin has a high level of two-way stretch and that differences in the percentages of skin stretch were small between men and women. The authors further studied the relationship between actual fabric horizontal stretch in wear and available fabric stretch measured at the seat of various garments while the subjects were in a sitting position, which is shown in Fig. 7-5. The skin strain was considerably higher than the actual garment stretch, indicating that garment fit and garment slip played an important role in accommodating the skin strain. Meanwhile, there was a direct relationship between available fabric stretch and the actual stretch. The higher the available stretch, the higher was the stretch in use. Also, the relationship between available and actual stretch varied with different types of garments, indicating the influence of the relative ratio between garment size and body size, and also the effect of body contact points.

Kirk and Ibrahim also investigated the relationship between the pressure on the body and fabric stretch level. The pressure, P, was calculated according the following equation:

$$P = T_H / \gamma_H + T_V / \gamma_V$$

where T is the tensile stress measure on the Instron at the same level of strain, and γ is the radius of curvature of the relevant body parts. Subscripts H and V indicate horizontal and vertical directions, respectively.

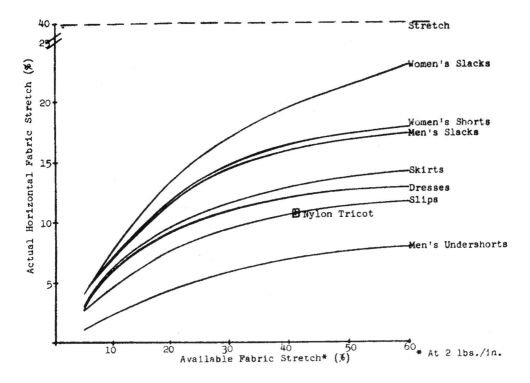

Fig. 7-5 Actual horizontal fabric stretch in wear, and available stretch [226]

Further, the authors studied consumer preference on stretch level in terms of comfort. It was found that higher stretch with lower power was always preferred, and that wearers' preferences for stretch were in a range from 25% to 45%, depending on the end use. Also, the direction of stretch relative to the body had significant impact on comfort. The results of this comfort study are summarized in Table 7-2.

Table 7-2
Comfort Preferences on Fabric Stretch [226]

End use	Direction of stretch	Preferred at least 75% of time	
		Stretch level (%)	Direction
Men's suit jackets	H	30	H
Men's suit slacks	H,V*	30	H
Men's slacks-casual	H	30	H
Men's shorts	H	30	H
Men's shirt-sleeves	H,V*	25	V
Men's shirt-body	H	25	H
Men's undershorts	H,V*	25	H
Women's slacks	H,V*	35	H
Women's shorts	H	35	H
Women's tensioned slacks	V	45	V
Skirts	H	25	H
Dresses	H	30	H
Slips	H	30	H

*End used where direction of stretch was compared. H = Horizontal, V = Vertical

Denton (1970) [227] pointed out that there are four mechanical factors relating to comfort, which are weight, ease of movement, stretch, and ventilation. The average weight of men's jackets, together with the contents in the pockets, was reported as about 1.5 kilos. This may contribute to discomfort perceptions as a relatively small area of the body usually supports it; the pressure generated by the weight on the skin may be above the comfort level. Ease of movement is largely dependent on garment design and the relative size between body and clothing. Loose fitting allows freedom of movement but may not be desirable in many situations. Tight fitting may be suitable for certain end-uses; however, it can exert pressure on localized areas of the body surface and cause discomfort. Ventilation normally occurs through openings at the two legs, two arms and the neck, and is influenced by garment fit and styling.

Stretch fabric, which can expand and contract without buckling or wrinkling, is another way to fit the body shape. Denton estimated the pressure discomfort threshold by stretching a band of elastic material around a part of the body and making a judgement of the level of comfort. It was found that, if the band was slightly uncomfortable initially, it became acceptable after a period, whereas if the band was very uncomfortable initially, it became intolerable as time passed. The pressure threshold of discomfort was found to be around 70 g cm^{-2}, which was close to the average capillary blood pressure of 80 g cm^{-2} near the skin surface. The pressure ranges of various types of garment reported by researchers are summarized in Table 7-3.

Table 7-3
Pressures Generated by Various Types of Clothing

Garment type	Pressure (g cm^{-2})	References
Swimwear	10–20	[228]
Modern corsets	30–50	[228]
Knitted foundations (modern girdle)	20–35	[228]
Elastic sock-tops	30–60	[228]
Medical stocking	30–60	[228]
Figure persuasive	< 20	[229]
Trousers braces	60	[227]

Growther (1985) [230] studied the comfort and fit of 100% cotton denim jeans. The author pointed out that classic jeans are characterized by body-hugging or tight fitting, which may result not only in a sculptured form, but also in possible body malfunction in the long-term. Rutten (1978) [231], an orthopedic surgeon, reported difficulties experienced by some patients who had worn non-stretch, tight-fitting jeans over an extended period. Other medical reports suggested that tight clothing could act as an effective tourniquet when the body assumed a sitting or crouching position, leading to thrombosis [232]. These researches suggested that the physical requirements of the wearers need to be satisfied by clothing in designing or developing apparel products. On the basis of a series of experiments, Growther concluded that the inherent properties of the fabric construction might be utilized to reduce skin strain and enhance body-contouring [230]. This may be achieved by adjusting the angle of the back-rise seam to one nearer to the true bias to meet skin stretch demands in the local hip-to-hip, local-crotch, and buttocks locations. Also, the front and back panels of the jeans can be pre-shaped to reduce restraint and provide local accommodation of body contours.

8. PREDICTABILITY OF CLOTHING COMFORT PERFORMANCE

8.1 Prediction of Fabric Hand

As discussed in the previous sections, the subjective perception of comfort of clothing by a wearer is determined by sophisticated psychological and physiological processes, which in turn are evoked by various physical stimuli. These physical stimuli are determined by a number of physical processes that are dependent on the relevant fiber–fabric–clothing physical properties and structural features. Therefore, it seems desirable and logical to develop methods to predict the comfort performance of clothing objectively. Considerable research work has been carried out to measure fabric properties and predict some aspects of clothing comfort performance through various approaches.

Evaluation of fabric hand through fabric objective measurement, which has been developed on the basis of the work of Kawabata and his co-workers, has widely been recognized and used around the world.

On the basis of the fundamental work on fabric mechanical properties and fabric hand, Kawabata (1973) [233,234] developed the KES-F system, which has been described in many publications [21,235–237]. The system uses four instruments manufactured by Kato Tekko Co., Kyoto, Japan:

KES-FB1   tensile and shear tester,
KES-FB2   bend tester,
KES-FB3   compression tester, and
KES-FB4   surface-friction and geometrical-roughness tester.

These instruments measure 15 parameters from fabric tensile to fabric surface roughness, as shown in Table 8-1.

Table 8-1
Testing Parameters of KES System

Fabric tensile:	EM	Fabric extension at 5 N/cm width
(KES-FB1)	LT	Linearity of load–extension curve
	WT	Energy in extending fabric to 5 N/cm
	RT	Tensile resilience
Fabric shear:	G	Shear rigidity
(KES-FB1)	2HG	Hysteresis of shear force at 0.5° of shear angle
	2HG5	Hysteresis of shear force at 5° of shear angle
Fabric bending:	B	Bending rigidity
(KES-FB2)	2HB	Hysteresis of bending moment
Fabric compression:	LC	Linearity of compression thickness curve
(KES-FB3)	WC	Energy in compression fabric under 5 kPa
	RC	Compressional resilience
	T_o	Fabric thickness at 50 Pa pressure
	T_m	Fabric thickness at 200 Pa pressure
Fabric friction and	MIU	Coefficient of steel–fabric friction
roughness:	MMD	Mean deviation of MIU
(KES-FB4)	SMD	Geometric roughness

Kawabata [233] used a stepwise linear regression procedure to develop his equation between subjective perception and the objective measurements for predicting primary hand values (HVs) in the form of linear and mixed linear–log functions. Hu *et al.* [135] examined the relationships between subjective primary hand values and the objectively measured fabric mechanical properties using the KES system in three forms: multivariate linear function, multivariate mixed linear–log function, and multivariate power function. They found that the multivariate power function gave much better predictions on three HVs than the other functions. Using step-wise regression, the authors found that HV1 (stiffness) was related to WC, B, MIU, MMD, and LC. HV2 (smoothness) was a function of T_o, SMD, B, LC, WT, W, HB, and WC. HV3 was dependent on T_o, MIU, RT, SMD, and WC.

8.2 Prediction of Clothing Thermophysiological Comfort

Fanger (1970) [4] developed a general thermal comfort equation that took account of physical activity level and clothing transport behavior, to calculate thermal comfort in all combinations of the environmental variables (air temperature, air humidity, mean radiant temperature, and relative air velocity). The equation was derived for creating optimal thermal comfort by detailed analysis of the influence of the individual variables, including a large number of comfort diagrams and their application in environmental engineering. In this equation, the human body was considered as a heat generator with its thermal balance achieved through various dry and latent heat transfer processes. The equation contained three sets of variables:

(i)   body activity, including metabolic rate of the body (M), the Dubois area (the surface area of the nude body, Adu), and the efficiency of external mechanical work (η);
(ii)   environmental variables, including air temperature, mean radiant temperature, pressure of water vapor pressure in ambient air, and air velocity;
(iii) function of clothing.

In describing the function of clothing, Fanger used two criteria: thermal resistance of the clothing and the ratio of the surface area of the clothed body to the surface area of the nude body.

For the purpose of development of an environmental temperature scale, Gagge *et al.* (1971) [5] developed a model that contained many physiologically dependent and independent variables. This model was aimed to predict the physiology of heat regulation and comfort which occurs during a thermal state of quasi-equilibrium after a fixed exposure period to various environmental conditions. The details of the model are discussed in Section 4.3. Compared with Fanger's equation, this model made significant improvement by taking into account the thermal regulatory mechanism of the human body in predicting thermal comfort. In this model, the functions of clothing were described by two criteria: F_{cl}, Burton's thermal efficiency factor measuring the efficiency for the passage of dry heat from the skin surface through the clothing to the environment; and F_{pcl}, Nishi's permeation efficiency factor for water vapor evaporated from the skin surface through clothing to the ambient air. These two parameters were analogous factors for heat transfer by convection and for mass transfer by water vapor transfer respectively. Gagge (1973) [144] reported an updated version of the model for the development of rational temperature indices of human beings' thermal environment. In this paper, the thermal efficiency factor, F_{cl} was given by the ratio $I_a/(I_a + I_{clo})$, where I_{clo} and I_a are the intrinsic insulation of the clothing worn and of air respectively. The permeation efficiency factor, F_{pcl} was defined as the ratio $h_{ecl}/(h_e + h_{ecl})$, where h_{ecl} is the intrinsic coefficient for permeation of water vapor through the clothing itself, and h_e is the evaporative heat transfer coefficient from the body surface to the environment. The values of these two parameters for normal everyday clothing were defined by the intrinsic clothing insulation in *clo* units as:

$$F_{cl} = 1 / (1 + 0.155\, h_t\, I_{clo})$$

$$F_{pcl} = 1 / (1 + 0.143\, h_{ch}\, I_{clo})$$

where h_{ch} is the convective heat transfer coefficient, $h_t (= h_r + h_{ch})$ is the combined heat transfer coefficient, and h_r is the linear radiation transfer coefficient.

In 1986, Gagge and his co-workers [6] reported a further improved model in the process of proposing a single standard index of comfort, health, and performance during rest and exercise, which combined the concepts of temperature and sensory indices developed over 60 years. In this improved model, the intrinsic clothing insulation (I_{cls}) was considered as a function of standard activity:

$$I_{clo} = 1.33/ (M - W + 0.74) - 0.095$$

where M and W (work) were in met units (1 met = 58.2 Wm^{-2}) and I_{clo} in *clo* units.

More recently, Li *et al.* (1998) interfaced Gagge's two-node model of a nude body with the model of coupled heat and moisture transfer of clothing discussed in Section 5, to describe the heat and moisture balance of a clothed body [157]. The major difference of this model to Fanger's and Gagge's models is that the clothing is not treated as a barrier with static thermal and moisture insulation values. Instead, the heat and moisture transport in clothing is treated as a dynamic and coupled process, the behavior of which is a function of fiber materials, body physical activities, and environmental conditions. The advantage of this combined model is that it enables us to examine the influence of the dynamic heat and moisture transfer process of clothing on the thermal regulatory responses of the body under various dynamic wear situations. With this model, it is possible to investigate the impact of different types of fibers and fabrics with different properties, on heat balance, thermoregulatory responses, and thermal comfort under various transient conditions.

Fig. 8-1 shows the temperature profiles from the external environment, through the clothing microclimate, to the core of the body during humidity transients in which theoretical predictions from the model were compared with experimental results. Comparing the predicted temperature profile for garments (see Fig. 4-6), the model predicted smaller skin and fabric changes with polyester than with wool during the humidity transients. The trends were essentially demonstrated by the experimental measurements, though the measured temperatures varied more than the predicted ones. Through the F-test, it was found that the differences in the measured skin and fabric temperature changes between wool and polyester were significant at 95% level during the period between 10 and 50 minutes. This suggests that the model is able to predict the influence of different fibers on the dynamic skin temperature responses of the body during humidity transient condition.

Comparison of theoretical predictions and experimental measurements on relative humidity in the clothing microclimate is shown in Fig. 8-2. Similar to the prediction for the wool garment in Fig. 4-7, the predicted microclimate relative humidity agreed with the experimental measurements.

Table 8-2
Umbach's Clothing Physiological Assessment System [238]

Level	Test methods	Explanation
1	Skin model	- biophysical analysis of textiles
		- thermophysiological prediction calculations for textiles
2	Thermal manikin	- biophysical analysis of clothing systems
3	Controlled wear trials with subjects in climatic chamber	- physiological body functional characteristics
		- subjective evaluations
4	Limited trial in practice	- limited field test by subjective assessment
5	Test market	- field test with large number of test subjects

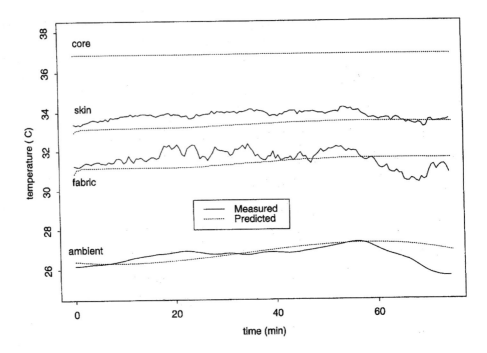

Fig. 8-1   Temperature profiles during humidity transients with polyester garment [157]

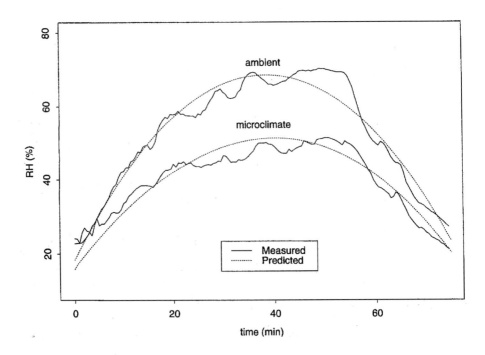

Fig. 8-2   Relative change of humidity in clothing microclimate during humidity transient with polyester garment [157]

Umbach (1987) [238] described an analytical system with five levels for assessing the physiological wearing properties of textiles and clothing, as summarized in Table 8-2.

At Level 1, fabric layers were tested for heat and moisture transfer behavior by using an artificial skin model that simulated the thermoregulatory process of human skin and its heat and moisture exchange with clothing. With this skin model, four sets of parameters could be measured, as shown in Table 8-3. These four sets of parameters were combined to derive formulae for predicting the physiological comfort of textiles. The formula for underwear was:

$$TK_T = \alpha_1 i_m + \alpha_2 F_l + \alpha_3 K_d + \alpha_4 \beta_T + \alpha_5 K_f + \beta$$

where TK_T is a comfort rating ranging from 1 (= excellent) to 6 (= insufficient). The constants for underwear were determined as: $\alpha_1 = -5.64$, $\alpha_2 = -0.375$, $\alpha_3 = -1.587$, $\alpha_4 = -4.512$, $\alpha_5 = -4.532$, and $\beta = 11.553$. The assumption in the development of this formula was that, during the period of wear, normal situations occur as frequently as those with moderate and heavy sweating.

At Level 2, a life-sized thermal manikin ('Charlie'), was used to determine the thermal insulation and moisture transport properties of the entire garment system, including fabrics used to make underwear and outerwear, and of the air layers in the clothing microclimate and at the external surface of the garment. The manikin was designed to simulate the topographic distribution of temperature on the surface of a human body for measuring the thermal insulation of the garment system (R_c). The manikin was also able to imitate various positions of the body, such as standing, walking, and lying to allow quantitative determination of the influence of convection and ventilation caused by the bodily motion of the wearer.

Table 8-3
Parameters Measured by the Skin Model [238]

No.	Testing parameters	Criteria calculated	Wear situation simulated
1	Heat transfer under stationary conditions	- thermal insulation R_{ct}	Normal stationary wear
2	Moisture vapor transfer under stationary conditions	- water vapor permeation resistance R_{et} - water vapor permeability index i_m	Normal stationary wear
3	Moisture vapor transfer under transient conditions	- moisture buffering index K_d - temperature buffering index β_T	Non-stationary wear – intermittent pulses of moderate sweating
4	Liquid transport under transient conditions	- liquid buffering index K_f - moisture permeability F_l	Non-stationary wear – intermittent pulses of heavy sweating

A set of thermal insulation values (in m^2 KW^{-1}) could be measured: $R_c(1)$ with a motionless manikin, $R_c(2)$ with a manikin moving at a defined speed and with sealed garment openings, and $R_c(3)$ with a manikin moving and wearing garments without sealed openings. The difference between $R_c(2)$ and $R_c(3)$ indicates the influence of ventilation in the garment system. These specific measurements for the complete garment system were used to develop a prediction model for showing the wearing properties under all possible climatic and activity conditions. The prediction model was integrated into a system of analysis. The definitions of the individual textile–clothing specific characteristics were obtained empirically by using regression equations derived from wear trials with test subjects in a climatic chamber. From the prediction model, a comfort rating, TK, which lies between 1 (= excellent) to 6 (= completely unsatisfactory), can be calculated from the following formulae:

Heat load for warm environment:

$$TK = 1.52 \ T_{re} + 0.3 \ T_s + 3.1 \ k_f + 0.02 \ HR - 767.85$$

Heat loss for cold environment:

$$TK = -5.61 \ \Delta T_{re} + 0.6 \ \Delta T_s + 1.51$$

where T_{re} is the rectal temperature; T_s is the mean skin temperature; k_f is the sweating covering rate of the skin; HR is the pulse frequency; and ΔT_{re} and ΔT_s are the decrease in rectal and mean skin temperature respectively.

At Level 3, controlled wear trials with test subjects were carried out in climatic chambers to test the predictions from the results obtained from the measurements and systems of analysis from Levels 1 and 2. At Levels 4 and 5, wear trials were carried out with a limited or large number of subjects in the field in which the garments are used. These types of wear trials were used mainly in fundamental studies for the particular types of clothing to determine the parameters that are important for the end uses.

This system of measurement and analysis has been applied to a wide range of functional designs of civilian work and protective clothing and uniforms, and to the development of test standards and application standards for various clothing. The system, however, focuses on thermophysiological comfort (i.e. the thermal–moisture comfort) of clothing, which is only one aspect of sensory comfort. Under normal wear situations, thermophysiological comfort is often found not to be the most important factor determining consumers' sensory comfort [37]. Therefore, there is a need to develop a more comprehensive understanding of the sensory comfort and objective measurement–prediction system.

8.3 Predictability of Sensory Comfort

From the fundamental investigation and modeling of the physical stimuli (heat and moisture interaction between skin and fabric), the neurophysiological responses, and subjective perceptions, it seems possible to predict the dynamic thermal comfort performance of clothing. As discussed in previous sections, the thermoregulatory responses and subjective perception of thermal and moisture sensations can be predicted by specification of relevant fiber properties, fabric structural features, and boundary conditions. With an understanding of the neurophysiological mechanisms of sensory perception of prickle and itch, the tactile comfort factor of clothing can be predicted from fiber-diameter distributions and fiber-end distribution densities at the fabric surface.

These research activities are fundamental and important for establishing the scientific framework to advance our knowledge of clothing comfort. However, the comfort of clothing involves a large number of complex and entangled factors. The overall sensory comfort performance of clothing is still difficult to predict. Therefore, it is worthwhile studying the predictability of and relationship between objective fabric properties and subjective sensory perceptions and preferences, by utilizing statistical methods.

Vollrath and Martin (1983) [239] compared the subjective judgment of skin-contact comfort in the back–shoulder–neck area with fabric properties such as weight, thickness, density, fiber fineness, friction coefficient, and compressibility. No fundamental relationship was found between fabric weight, thickness, or density, and the subjective skin contact sensation. No correlation was observed between fiber fineness, friction coefficient, or compressibility and comfort statements. Drape and bending rigidity were weakly correlated. Surface roughness values obtained by an electromechanical means were strongly correlated. They concluded that tests with reliable and sensitive test personnel could not be replaced by laboratory-apparatus tests.

In an attempt to establish the relationship between subjective comfort perceptions and fabric physical properties, Li (1988) [220] carried out a series of psycho-physiological wear trials using T-shirts made from 8 types of fibers. In the wear trials, subjective ratings on 19 sensory descriptors were recorded under two environmental conditions [37], from which three fundamental sensory factors were identified as *thermal-wet, tactile, and pressure comfort* [41,65].

On the other hand, an extensive range of fabric physical properties was tested as shown in Table 8-4, from which 42 objective criteria were obtained. The properties measured were classified into two categories: transport properties and mechanical properties. Transport properties included thermal resistance, water vapor and air permeability, demand wettability, drop wettability, and water evaporation propensity, with 14 criteria calculated. By principal component analysis, five independent factors were extracted from the 14 transport criteria, which indicate fabric wettability, permeability, water evaporation propensity, and resistance to water absorbency and water vapor. Also, principal component analysis was used to extract five factors from the 28 mechanical criteria, which were the roughness and fullness, stiffness, perpendicular deformation of the fabrics, tensile stiffness of the fibers, and tensile stiffness of the yarns.

Table 8-4
Summary of Measurement Methods on Fabric Properties [38]

Thermal resistance	B.S. 4745
Air permeability	B.S. 5636
Water vapor permeability	Dish method with and without temperature gradient
Wettability	Drop absorption rate
Demand wettability	Porous absorption rate
Water evaporation propensity	Rate of evaporation from initially saturated sample under standard environmental conditions
Tensile tests (fiber and yarn)	B.S. 1932, B.S. 3441
Tensile tests (fabric)	Cyclic deformation, wale, and course directions
Bending test	Loop bending, wale, and course directions
Plate compression test	Parallel plate compression
Compression cage test (bagging)	Cyclic compression test by applying load normal to the surface of the fabric mounted on the horizontal open face of a metal cage
Fiction coefficient	Force generated by moving surfaces of two identical fabrics in contact at standard rate and pressure
Surface roughness	Force generated by moving thin wire over surface at standard rate and pressure

Using canonical correlation analysis, Li [220] studied the predictability between the ten physical factors and the three psychological sensory factors. It was found that the sensory factors were significantly related to the corresponding dimensions of the physical properties of the fabrics. The *tactile* comfort factor was mainly related to fabric roughness and fullness, fabric stiffness and wettability. The *pressure* comfort factor was closely correlated to fabric stiffness, fabric permeability, and fiber tensile stiffness. The *thermal-wet* comfort factor was related to fabric wettability, fabric roughness and fullness, and fabric water evaporation propensity.

Canonical redundancy analysis showed that the canonical variables of the physical properties of fabrics were reasonably good predictors for the psychological sensory factors, with a cumulative redundancy over 0.71. The sensory factors, however, were poor predictors for the physical properties of fabrics, with a cumulative canonical redundancy of less than 0.376. Therefore, it appeared that objective measurements of a wide range of fabric physical properties were able to predict the sensory comfort of clothing worn next to the skin, reasonably well.

8.4 Predictability of Subjective Preferences

Li *et al.* [38] also applied canonical correlation and redundancy analysis to investigate the predictability of subjective preferences from the objective physical factors of fabrics. Three sets of subjective preferences towards 8 sets of T-shirts were obtained by asking the subjects to judge their comfort performance by handling them and then by wearing them under two environmental conditions. Two complete paired comparison experimental runs were carried out, with 56 observations in each condition. Then, the preferences were analyzed using Thurstone's comparative judgment method. The derived preference votes were compared with the ten objective physical factors by canonical correlation analysis.

Three significant canonical correlation coefficients were obtained, indicating that there were three dimensions of the objective physical factors related to the subjective preference votes. The first canonical correlation showed that the subjective wearing preference votes were closely associated with fabric roughness and fullness, fabric wettability, and fabric perpendicular deformability. The second suggested that the handling preference votes were mainly related to fabric stiffness, fabric perpendicular deformability and yarn stiffness, as well as fabric wettability. The third indicated that the wearing preference votes were also related to fabric wettability, fabric roughness and fullness, and tensile stiffness of fibers.

From canonical redundancy analysis it was found that the objective factors of fabrics had great predictive power for the subjective preference votes, with cumulative proportion of variance 0.983. This was further confirmed by the squared multiple correlation, which showed that the first three canonical variables of the objective physical factors of fabrics had very good predictive power for all the three subjective preference votes. The authors concluded that objective laboratory measurements of physical properties of fabrics showed good ability to predict the subjective preferences for clothing, provided that enough information about the physical behavior of the fabric was obtained.

9. APPLICATION OF CLOTHING COMFORT RESEARCH

9.1 Industrial Applications

From the discussions in previous sections, it is obvious that significant progress has been made in various clothing comfort research areas. This progress has great potential for application in the textile and clothing industry. In modern consumer markets, to meet and exceed customer expectation is the key to success for any enterprise. As consumers in the world demand higher comfort performance from apparel products and more technical information, comfort research is becoming more and more important for individual companies in the industry. This has been demonstrated clearly from the successful comeback of the synthetic fiber industry through development of high-tech apparel products with emphasis on comfort and functional performance.

The knowledge and methodology developed in clothing comfort research can be applied in a number of ways:

- to conduct consumer research by utilizing the research techniques developed to understand what consumers want and need, and to identify a market gap for new product development;
- to develop textile products that have unique functional features and are sure to satisfy targeted customers, by using the technical knowledge obtained;
- to use consumer sensory evaluation as a way for new product evaluation to reduce the risk of market failure;

- to develop technical information and specifications for promotional and marketing purposes; and
- to formulate quality-control tools by developing test methods, instruments, and standards.

9.2 Consumer Research

Understanding and predicting human needs and behavior is one of the most important methods for modern enterprises in making sound business strategies to meet consumers' requirements and combat market competition. Key business strategies involve brand management, new product development, and market development. Consumer research needs to identify buyers' product-related attitudes and find why markets change, in order to provide good estimations of changes to sales volumes. Through such attitude studies, all the criteria affecting buyers' decisions between brands can be identified and screened. Typically, hundreds of criteria or attributes, including both functional and emotive criteria, are generated, and 10–15 possible underlying dimensions of choices are identified, from which individual consumers use two to five criteria to make their purchase decisions [240].

Creig (1994) [240] pointed out the importance of systematically studying product attributes, and their relative importance in motivating buyers, by inferential procedures. A set of functional and emotional product attributes needs to be identified for explaining buyers' brand preferences among all the brands or choices. Through this type of analysis, a picture can be built up to show which brands are competing on what attributes and where the market share is most likely to be profitably increased in regard to the ease of changing some images over others.

However, it is one of the most difficult tasks in social science to obtain an understanding of consumers' needs and wants in order to predict their purchase behavior. Without a good understanding of consumers, low success rates of new product launches and the waste of half of advertising expenditure have been observed traditionally [240]. Dayal (1980) [241] reported that consumers are not adept at identifying what are the appealing qualities in a garment, while Greenwood [242] pointed out that consumers are generally not required to describe the characteristics which they seek in purchasing garments in any detail. This suggests that consumers' vocabulary for garment attributes is underdeveloped. Creig [240] argued that the relative importance of the product attributes should not be asked directly of buyers. A reasonable ranking of functional importance can be derived by direct questions, but not of emotional attributes. Consumers usually underestimate the importance of emotional attributes and overestimate the attributes related to social and political issues.

During the process of purchase decision making, consumers evaluate the appeal of the garments as a whole [243]. The process involves many individual factors and their integration includes the physical properties of the clothing, physiological characteristics of the buyers, their wearing experiences, and economic and social background [9,37,244,245]. As discussed in the previous sections on evaluation of clothing sensory comfort, a complex variety of stimuli stimulate various sensory perceptions. The judgment of the overall appeal of a garment comes from a complicated experience involving the integration of individual sensations. Many of the sensations cannot be experienced without exposure to a specific combination of physical activities, environmental conditions, and the physiological–psychological status of the wearers. The traditional market research methods and consumer surveys can derive information on brands that is of a very general nature, but cannot provide a more detailed understanding of the roles of physical–mechanical attributes in consumer desirability of fabrics and garments. Consumers are generally not required, or are unable, to pinpoint appealing properties of garments for specific end-uses without specific references. Therefore, more advanced or more specific consumer research techniques are required to obtain consumer information for product development, brand management, and development of market strategies.

Creig [240] pointed out that by using the new generation of consumer research tools, business firms could select product development, new product strategies, and rational–emotional advertising messages with more accuracy. Employing such techniques in several companies, the predictions of brand market-share changes were achieved within 10% of the actual for 80% of the time, and 80% of new product success rates was observed.

The techniques developed over the years in clothing comfort research and consumer sensory perceptions can become powerful tools for these purposes. The techniques developed by Hollies [246] for generating individual descriptors for sensations experienced during wearing of clothing by human subjects can be used for obtaining information on product attributes desired by consumers. The wear trial techniques can be effective means to obtain more specific and accurate sensory responses from consumers with fresh perceptual experiences on clothing in controlled dynamic ambient temperature and humidity changes. The personal construct theory developed by Kelly [23] validates these approaches theoretically on the basis that human beings have the ability to be specific and draw on internal concepts of a particular type of garment for generating specific criteria to evaluate the garment. The semantic differential scaling techniques that have been developed and applied successfully in studies of fabric hand and clothing comfort by many researchers such as Winakor (1980) [50], David (1985) [20], and Fritz (1992) [52], can be effective in identifying consumers' desired product attribute profiles. The three major underlying sensory dimensions in consumers' comfort perceptions, which are identified from a large number of sensory studies [41], can be used as the criteria for evaluating clothing comfort performance during wear. The relative contributions of the three comfort dimensions to consumers' overall preferences or sensory responses [66], which are obtained by analyzing the results from various sensory studies, can be useful guidelines for development of apparel products for many specific combinations of physical activities and environmental conditions.

Therefore, application of these techniques, and integration of them into the conventional marketing research methodology, can become a powerful consumer research tool for companies in the textile and clothing industry, for product development, and for marketing management. In summary, the sensory study techniques developed in clothing comfort have two major advantages:

(i)   Using sensory study techniques, we can more accurately identify the specific product attributes that consumers want to have, the attributes on which certain brands' competitive edges are based, and the market gap that can be filled by developing and positioning new product lines;

(ii)   Conducting focus group study or wear trials in simulated practical wear situations, we are able to obtain much more accurate sensory responses and preferences from consumers as they can draw their conclusions and make judgments on the basis of current experiences instead of vague memories of past experiences.

9.3 New Product Development
In modern consumer markets, companies that fail to develop new products are facing great risk. Their existing products are vulnerable to various changes in the market places, such as changes in consumer needs and preferences, new technologies, shortened product-life cycles, and increasing competition from substitutes and foreign–domestic companies. Meanwhile, new product development is risky with a disturbing failing rate. It has been estimated that 75 to 80%

of new products fail at launch [247,248]. Cooper and Kleinschmidt [249] studied the common features of successful new products across a wide range of industries and concluded that the number one success factor is a unique, superior product with higher quality, new features, and higher value in use. New products with a high advantage succeed 98% of the time, while new products with a moderate advantage have 58% success, and those with minimal advantage have only 18% success. The second success factor is a well-defined product concept prior to development, where the target market, product requirements, and benefits are carefully defined and assessed. Other factors include the synergy between technology and marketing, and market attractiveness. These findings illustrate the importance of understanding consumers' requirements and of using new technology to develop products unique to the end-users to ensure their success.

Clothing markets in the 90s are highly competitive. Consumers are demanding more functions, higher quality, and more added value. Fashion is no longer dictated by a handful of couture designers. Instead, consumer lifestyles are having a growing impact on the direction of fashion trends. Today, consumers want to dress more casually and can do so, as more and more companies institute less formal dress policies. At the same time, consumers desire garments that enable them to be comfortable and feel good in, whatever activity they happen to be engaged upon. They want their clothing to be natural, comfortable and 'easy care', and demand more information about the products that they buy. With these changes, development of new products to satisfy consumers' needs and wants has become increasingly more difficult and more important for competing in the consumer markets.

To increase the success rate of new products, it is important to find an effective way to identify the target consumers' specific requirements, and link them with the technical attributes of products. The knowledge and methodology developed in clothing comfort research are the exact weapons that are needed to achieve this objective. Psychological sensory study identifies what consumers want and desire, which is linked to the technical features (or physical stimuli) of products through psychophysical perceptual trials. Further, the physical mechanisms that generates such technical features are able to provide us with the technical specification and guidelines (know-how) to produce the desired products.

In summarization of the psychological–psychophysical–physical processes, a system for apparel product development and evaluation, which consists of five levels, is shown in Fig. 9-1. At Level 1, traditional consumer research techniques, such as focus groups, personal interviews, and consumer surveys, are carried out to obtain market information on the products. Information obtained in these types of research is general in nature with limited accuracy, as there are many uncontrollable variables in environment, human respondents, and market situations. At Level 2, wear trials can be conducted with a limited or large number of subjects in the field in which the garments are used. These types of wear trials are mainly used in consumer evaluation, market testing, or in fundamental studies for particular types of clothing, to derive more specific information on product attributes required by the end-uses. This type of research is expensive and has relatively high risk as some of the confidential information may be detected by competitors.

At Level 3, controlled wear trials with human subjects are carried out in climatic chambers for psychological sensory study, consumer focus group study, and subjective evaluation of apparel products. Very specific information on desired product attributes can be obtained from consumers' sensory responses. This type of research is more accurate and safe, as the variations in environmental conditions and human subjects are controlled with a certain degree of accuracy in the laboratory. Also, the psychological assessment can be used to test the predictions or results obtained from the measurements and systems of analysis from Levels 4 and 5.

At Level 4, the entire garment system is tested on human subjects or thermal manikins to determine the thermal insulation and moisture transport properties, mechanical behavior, and appearance during wear. These physiological and physical measurements are aimed to determine and evaluate the technical details of apparel products. The technical information needed includes the design and style of garments, the effect of body movements and environmental conditions, and the effect of garment assemblies from underwear and outwear to air layers in clothing microclimate and at the external surface of the garment. Research at this level can provide the essential technical specifications for the garment design and manufacturing processes.

At Level 5, fabrics are tested for a range of physical properties, from the heat and moisture transfer behavior, and mechanical behavior, to color and surface features. These measurements are used to evaluate the features and quality of fabric materials for achieving the desired comfort and functional performance during wear. The information obtained here can provide the detailed technical specifications for the textile processing and manufacturing processes.

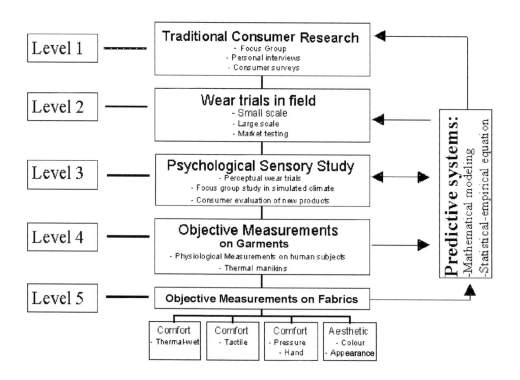

Fig. 9-1   System for clothing comfort and functional design evaluation

From Level 1 to Level 5, the information flows from market requirements to technical specifications; from being general to being specific. Using such a system, new products can be developed according to market demand and can satisfy the requirements of consumers on various product attributes such as comfort, easy care, and appearance maintenance.

On the other hand, from Level 5 to Level 1, information flows from being technical to being sensory and subjective. This means that, utilizing the knowledge and understanding of the physical mechanisms, psychophysical mechanisms, and psychological preferences of consumers on various sensory perceptions, we can develop predictive tools using either statistical or mathematical models for product development purposes. With these tools, manufacturers can conveniently optimize the product design according to the identified consumers' requirements prior to actual production. This enables textile and clothing manufacturers to save money and time, and to respond to market demand quickly and with confidence.

Umbach (1981) [250] reported a model for clothing comfort and functional design using, as an example, a cold protection garment which was an optimal multi-layered garment with controllable ventilation. With an optimal combination of fabrics and garment design, the protective clothing was able to maintain a comfort able microclimate in a wide range of external conditions from −20°C to 16°C, which was confirmed by tests using a thermal manikin and human subjects in climatic chambers. Similar approaches were applied to leisure and sportswear such as ski-suits, motorbike and sailing suits, and all-weather coats. Great market success was reported [238].

9.4 Consumer Subjective Evaluation

In new product development processes, expenses increase with the progressing stages. The ratios of costs on one successful new product were 1:5:25:31:156 from idea screening, concept testing, product development, and test marketing, to final market launch [247]. The most expensive cost is located at the final stage — the market launch. Therefore, it is extremely important to ensure that consumers' perceptions and acceptance of the new products are positive before launching into the market. To ensure this, consumer subjective evaluation of the new products must be carried out at the various stages. For apparel products, this is extremely important because many sensations cannot be perceived without wearing the garments in the actual wear situations that are determined by the combinations of physical activities, environmental conditions, and social environment.

The techniques of sensory studies described in the system shown in Fig. 9-1 can be utilized effectively to obtain consumers' responses at the stages of idea generation, idea screening, concept testing and product development in the form of functional tests, and consumer tests. The tests can be carried out at three levels: climatic chamber wear trials, limited field tests, and large-scale field tests. The functional tests can be carried out in the laboratory or field to ensure that the products perform safely and effectively through a range of physical and physiological tests at Level 4 of the system. Consumer tests can be conducted through a number of ways, from bringing consumers into a climatic chamber for simulated wear trials to conducting field wear trials. A range of techniques of sensory study, such as simple ranking, paired comparison, rating scales, and semantic scales, can be used to obtain consumers' perceptions and preferences. Abundant information can be obtained through such tests, such as the perceived superiority of the products, the perceived difference between consumers' ideal products and the developed products, the relative positions of the newly developed products with own and competitors' existing products, and the perceived value and price of the new products. This information is very critical for companies to make important business decisions, such as *go* or *drop*, product modification, branding and positioning, packaging, distribution, pricing, and other market launching strategies such as promotion and advertising.

9.5 Quality Control and Troubleshooting

The objective test methods developed in clothing comfort research can be used to derive a quantitative measure of the clothing comfort-related properties of fabrics and garments and to develop comfort specifications. Umbach [238] reported that the physiological specifications developed from his measurement system have been included in some German Industry DIN standards for work and protective clothing, as shown in Table 9-1. For instance, the weather-protective suit that is used in building construction for protection against the influence of wetness and wind, is required to have a water vapor resistance of not greater than 0.200 m²mbar/W according to DIN 61 539. This is aimed to maintain sufficient comfort for the wearers who work with moderately heavy physical activity at ambient temperatures up to 20°C. According to DIN 61 537, the values of fabric thermal resistance and water vapor resistance for protective vests against cold were specified for the purpose of extending the range of use of the weather protective clothing defined by DIN 61 539 down to −5°C.

These standards, based on parameters measured on clothing wear comfort performance, have been used as the technical specifications for industrial clothing and military uniforms. The techniques have also been applied to the areas of leisure, sport, and everyday clothing, such as technical specifications for weather-protective clothing, ski suits, anoraks, motorbike and sailing suits, all-weather coats, sports underwear such as ski underwear, and football and tennis tops. When the products are designed according to the technical specifications and produced in fashionable colors, fashion and function are harmonized and work together to make a better chance of market success [238].

Table 9-1
Some DIN Standards for Work and Protective Clothing

Test standard	DIN 54 101 (T 1 (E))	Skin model
Application standards	DIN 61 539	Weather protective suit
	DIN 61 537 (E)	Protective vest against cold
	DIN 30 711 T3	Warning clothing
	DIN 30 711 T 2	Warning clothing
	DIN 32 763	Protective clothing against chemicals Type 2

The work carried out by Umbach and his colleagues has demonstrated the effective application of research on clothing thermophysiology and comfort that has achieved significant market success. However, clothing comfort has much broader dimensions than thermophysiological comfort. As discussed in Section 3, studying the pattern of relationships among the sensory responses during wear shows that the comfort of clothing has three independent sensory factors: *thermal-wet comfort, tactile comfort, and pressure (body-fit) comfort*[41]. Further, from analysis of the sensory responses obtained from the survey and wear trials, it was found that the relative importance of the three factors in consumers' sensory experiences varies with different combinations of physical activities and environmental conditions, as shown in Fig. 9-2. For winter wear, the *tactile* and *thermal-wet* comforts were perceived to be important, taking about 67% of the total variance. For summer wear and spring wear, the *thermal-wet* comfort was perceived to be important, followed by *tactile* and *pressure* comforts. However, these three factors only account for about 60% of the total variances. For sportswear, the *thermal-wet* comfort was perceived to be extremely important, accounting for 43% of the total variances, followed by *tactile* and *pressure* comforts. It should be noted that these results have a basic assumption that the style and appearance of the garments are the same or meet the standards required by the consumer [252].

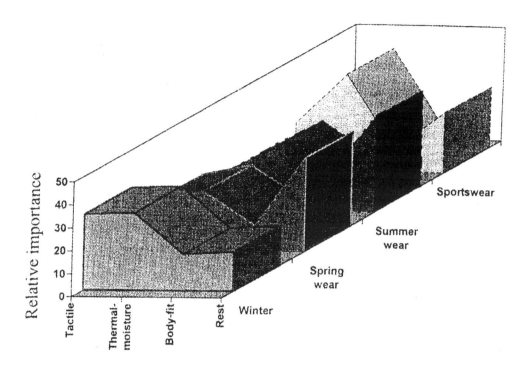

Fig. 9-2 Consumers' sensory requirements [252]

On the other hand, consumers' requirements for apparel products are for much more than comfort. Fig. 9-3 shows some results from a study on the perceptions of the importance of various attributes of clothing worn at the work place, which covered a number of national groups from Europe, Asia, and Australia. As shown, the main trends among the national groups were similar. From the nine dimensions listed, comfort, fit, style, color, and quality were rated more important than the others, which suggests the need for an effective method for evaluation of the overall wear performance of clothing. Vatsala and Subramaniam (1993) [251] proposed a product-performance index for integral evaluation of fabric performance, which is based on giving weightages for four functions: comfort, aesthetic appeal, ease of care, and durability.

A comprehensive system for overall evaluation of apparel products can be crucial for new product success. As all the final judgments are made by consumers on the basis of their perceptions and experiences during practical wear situation, the system described in Fig. 9-1 can be used for overall evaluation of apparel products. With such a system, the techniques developed in clothing comfort research can be used effectively to improve the quality of life for modern and future consumers. Also, utilization of the knowledge and methodology can become powerful tools for companies in the textile and clothing industry to develop product-based business strategies and achieve sustainable competitive advantages in the market place.

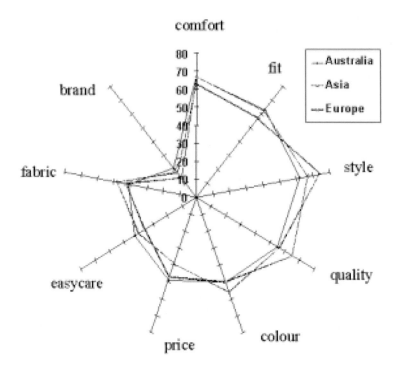

Fig. 9-3   Consumers' requirements of clothing [252]

NOMENCLATURE

A human body surface area (m^2);

AT_{ab} adjusted dry bulb temperature (°C);

C convective heat loss from the body ($W\ m^{-2}$);

C_a water vapor concentration in the air filling the inter-fiber void space ($kg\ m^{-3}$);

C_{ab} water vapor concentration in the ambient air ($kg\ m^{-3}$);

c_{bl} specific heat of blood = 1.163 ($kJ\ kg^{-1}\ K^{-1}$);

C_{cf} water vapor concentration at the fiber surface ($kg\ m^{-3}$);

C_f water vapor concentration in the fibers ($kg\ m^{-3}$);

C_{fi} water vapor concentration at the inner surface of the fabric ($kg\ m^{-3}$);

C_{res} heat lost by convection from the lungs during respiration ($W\ m^{-2}$);

C_{sk} water vapor concentration at the skin surface ($kg\ m^{-3}$);

C_v volumetric heat capacity of the fabric ($kJ\ m^{-3}\ K^{-1}$);

D_a diffusion coefficient of water vapor in air ($m^2\ s^{-1}$);

D_f diffusion coefficient of water vapor in the fiber ($m^2\ s^{-1}$);

D_{sk} water vapor diffusion coefficient of skin ($m^2\ s^{-1}$);

E Young's modulus of the fiber;

E_{comf} evaporative heat loss by sweating occurring in a state of comfort ($W\ m^{-2}$);

E_{dif} heat lost by water vapor diffusing through the skin layer ($W\ m^{-2}$);

E_{res} heat lost by moisture evaporation from the lungs during respiration ($W\ m^{-2}$);

E_{rsw} heat lost by sweat evaporation during body temperature regulation ($W\ m^{-2}$);

E_{sk} average latent heat flux from the skin ($W\ m^{-2}$);

F_{cl} Burton's thermal efficiency factor;

F_{pcl} Nishi's permeation efficiency factor for water vapor evaporated from skin surface through clothing to the ambient air;

H_a fractional relative humidity of the air;

h_c convective mass transfer coefficient at the outer boundary air layer ($m\ s^{-1}$);

h_{cf} mass transfer coefficient at fiber surface ($m\ s^{-1}$);

h_{ci} convective mass transfer coefficient in the clothing microclimate ($m\ s^{-1}$);

h_{ct} convective heat transfer coefficient at the outer boundary air layer ($W\ m^{-2}\ K^{-1}$);

H_f fractional relative humidity at the fiber surface;

h_t combined (convection and radiation) heat transfer coefficient at outer boundary air layer ($W\ m^{-2}\ K^{-1}$);

h_{ti} combined heat transfer coefficient in the microclimate ($W\ m^{-2}\ K^{-1}$);

I moment of inertia, $= \pi d^4/64$ in the case of a circular rod;

I_{clo} intrinsic clothing insulation (clo. 1 clo = $0.155 m^2\ kW^{-1}$);

i_m permeability index of clothing;

K thermal conductivity of the fabric ($W\ m^{-1}\ K^{-1}$);

K_d dynamic differential sensitivity associated with temperature change (impulses K^{-1});

K_{min} minimum thermal conductance of the skin tissue in the absence of skin blood flow = 5.28 ($W\ m^{-2}\ K^{-1}$);

K_s static differential sensitivity related to the steady-state temperature (impulses $K^{-1}\ s^{-1}$);

l length of the protruding fiber ends (m);

L thickness of the fabric (m);

L_{sk} thickness of the outermost layer of the skin (m);

m body mass (kg);

M metabolic heat production (W m^{-2});

m_{cr} body core mass (kg);

M_d moisture flow into clothing (kg m^{-2} s^{-1});

MRT mean radiant temperature (°C);

m_{sk} skin mass (kg);

M_t dry heat flow into clothing (W m^{-2});

p proportion of moisture uptake by the fibers during the second stage of sorption (fraction);

P_a ambient vapor pressure (mm Hg);

P_E critical buckling load of the protruding fiber ends (N);

p_h proportion of dry heat lost from the skin which passes through the inner surface of the clothing (fraction);

p_m proportion of moisture vapor lost from the skin which passes through the inner surface of the clothing (fraction);

P_{sk} saturated water vapor pressure at skin temperature (mm Hg);

$Q(y,t)$ pulse output response of a thermoreceptor as a function of time t and the depth y of the thermoreceptor in the skin (inpulses s^{-1});

r radial distance from the center of the fiber (m);

R radiant heat loss from the skin (W m^{-2});

R_1 first-stage moisture sorption rate of the fiber (kg m^{-3} s^{-1});

R_2 second-stage moisture sorption rate of the fiber (kg m^{-3} s^{-1});

R_c heat transfer resistance of clothing (Km^2 W^{-1});

R_e moisture transfer resistance of clothing (mm Hg m^2 W^{-1});

R_f mean radius of the fibers (m);

R_s sensation magnitude;

S heat storage of the body (W m^{-2});

S_{cr} heat storage of the core of the body (W m^{-2});

S_p physical stimulus magnitude;

S_{sk} heat storage of the skin of the body (W m^{-2});

S_v specific volume of the fabric (m^{-1});

t real time (s);

T temperature of the fabric (°C);

T_{ab} temperature of the ambient air (°C);

T_{cr} temperature of the body core (°C);

T_{fi} temperature at the inner surface of the fabric (°C);

T_i internal body temperature (°C);

T_k comfort rating ranging from 1 = (excellent) to 6 = (completely unsatisfactory);

T_{kt} comfort rating ranging from 1 = (excellent) to 6 = (insufficient);

T_{mb} mean body temperature (°C);

T_{mbo} initial mean body temperature (°C);

T_{sk} temperature of the skin (°C);

TC_{cr} thermal capacity of the body core (W hr K^{-1});

TC_{sk} thermal capacity of the skin (W hr K^{-1});

V_{bl} rate of skin blood flow (liter hr^{-1} m^{-2});

W work accomplished (W m^{-2});

W_c fractional water content at the fiber surface;

x distance from inner surface of fabric (m);

Y   thermal sensation on the scale of 1 = (cold), 2 = (cool), 3 = (slightly cold), 4 = (comfortable), 5 = (slightly warm), 6 = (warm), and 7 = (hot).

Greek symbols

ε porosity of the fabric (fraction);

λ heat of sorption of water vapor by fibers (kJ kg^{-1});

æ actual ratio of skin to total body mass;

τ effective tortuosity of the fabric (ratio);

ρ density of fiber (kg m^{-3}).

REFERENCES

[1]   Anon. Fibers of the Nineties. *Textile View Magazine*, 1991, 1–11.

[2]   L. Fourt and N.R.S. Hollies. *Clothing: Comfort and Function*, Marcel Dekker Inc., New York, NY, USA, 1970.

[3]   N.R.S. Hollies and R.F. Goldman. *Clothing Comfort: Interaction of Thermal, Ventilation, Construction, and Assessment Factors*, Michigan, Ann Arbor Science Publishers Inc., Ann Arbor, USA, 1977.

[4]   P.O. Fanger. *Thermal Comfort*, Danish Technical Press, Copenhagen, Denmark, 1970.

[5]   A.P. Gagge, J.A.J. Stolwijk, and Y. Nishi. An Effective Temperature Scale based on a Simple Model of Human Physiological Regulatory Response. *ASHRAE Trans.*, 1971, **77**, 247–262.

[6]   A.P. Gagge, A. Fobelets, and L.G. Berglund. A Standard Predictive Index of Human Response to the Thermal Environment. *ASHRAE Trans.*, 1986, vol. 2B.

[7]   K. Slater. *Human Comfort*, Thomas Sprinfield, USA, 1985.

[8]   K.L. Hatch. *Textile Science*, West Publishing Company, New York, NY, USA, 1993.

[9]   G.J. Pontrelli. Partial Analysis of Comfort's Gestalt, in *Clothing Comfort* (eds N.R.S. Hollies and R.F. Goldman), Ann Arbor Science Publishers Inc., Michigan, USA, 1977, 71–80.

[10]   K. Slater. The Assessment of Comfort. *J. Text. Inst.*, 1986, **77**, 157–171.

[11]   F.A. Geldard. *The Human Senses*, New York: John Wiley & Sons Inc., 1972.

[12]   P. Dunn-Ranking, *Scaling Methods*, Lawrance Erlbaum Associates Publishers, London, UK, 1983.

[13]   N.R.S. Hollies. Psychological Scaling in Comfort Assessment, in *Clothing Comfort* (eds N.R.S. Hollies and R.F. Goldman), Ann Arbor Science Publishers Inc., 1977, Michigan, USA, 107–120.

[14]   N.R.S. Hollies. *Investigation of the Factors Influencing Comfort in Cotton Apparel Fabrics*, US Department of Agriculture, New Orleans, USA, Contract 12-14-100-7183 (72), 1965.

[15]   N.R.S. Hollies, A.G. Custer, C.J. Morin, and M.E. Howard. A Human Perception Analysis Approach to Clothing Comfort. *Text. Res. J.*, 1979, **49**, 557–564.

[16]   N.R.S. Hollies. The Comfort Characteristics of Next-to-skin Garments, Including Shirts, presented at *3rd Shirley Int. Seminar*, Manchester, UK, 1971.

[17]   N.R.S. Hollies. Improved Comfort Polyester. Part IV: Analysis of the Four Wear Trials. *Text. Res. J.*, 1984, **54**, 544–548.

[18]   W.S. Howorth and P.H. Oliver. The Application of Multiple Factor Analysis to the Assessment of Fabric Handle. *J. Text. Inst.*, 1958, **49**, T540.

[19]   W.S. Howorth. The Handle of Suiting, Lingerie, and Dress Fabric. *J. Text. Inst.*, 1964, **55**, T251–260.

[20]   H.G. David, A.E. Stearn, and E.F. Denby. The Subjective Assessment of Handle, presented at *Proc. of 3rd Japan-Australia Symp. on Objective Measurement: Applications to Product Design and Process Control*, Kyoto, Japan, 1985.

[21]   S. Kawabata and M. Niwa. Fabric Performance in Clothing and Clothing Manufacture. *J. Text. Inst.*, 1989, **80**, 19.

[22]   Y. Li. Sensory Comfort: Consumer Responses in Three Countries. Accepted by *J. China Text. University.*

[23]   G.A. Kelly. Man's Construction of his Alternatives, in *Assessment of Human Motives* (ed. by G. Lindzey), Holt, Rinehart & Winston, New York, NY, USA, 1958.

[24]   A.M. Fritz. Sensory Assessment Assessed. *Textile Asia*, 1990, **21**, 144–147.

[25]   D. Laming. *Psychophysics*, in *Sensation and Perception* (eds R.L. Gregory and A.M. Colman), Longman, London, UK, 1996, 97–123.

[26]   T. Engen. *Psychophysics*, in *Sensory Systems II: Senses Other than Vision* (ed. J.M. Wolfe), A Pro Scientia Viva Title, Boston, USA, 1988, 104–106.

[27]   D.S. Tull and D.I. Hawkins. *Marketing Research: Measurement and Method*. Macmillan Publishing Company, New York, NY, USA, 1993.

[28]   S.S. Stevens. On the Theory of Scales of Measurement. *Science*, 1946, 677–680.

[29]   M. Traylor. Ordinal and Interval Scaling. *J. Market Res. Soc.*, 1983, **25**, 297–303.

[30]   M.R. Crask and R.J. Fox. An Exploration of the Internal Properties of Three Commonly Used Research Scales. *J. Market Res. Soc.*, 1987, **29**, 317–339.

[31]   G.R. Dowling and D.F. Midgly. Using Rank Values as an Interval Scale. *Psychology and Marketing Research*, 1991, 37–41.

[32]   Y. Li, B.V. Holcombe, and F. Apcar. Moisture Buffering Behaviour of Hygroscopic Fabric During Wear. *Text. Res. J.*, 1992, **62**, 619–627.

[33]   Y. Li, A.M. Plante, and B.V. Holcombe. The Physical Mechanism of the Perception Dampness in Fabrics. *J. Thermal Biology*, 1993, **18**, 419–419.

[34]   A.M. Plante, B.V. Holcombe, and L.G. Stephens. Fiber Hygroscopicity and Perception of Dampness. Part I: Subjective Trials. *Text. Res. J.*, 1995, **65**, 292–298.

[35]   P.E. Green, D.S. Tull, and G. Albaum. *Research for Marketing Decisions*, Prentice-Hall Inc., Englewood Cliffs, USA, 1988.

[36]   J.F. Fuzek and R.L. Ammons. Techniques for the Subjective Assessment of Comfort in Fabrics and Garments, in *Clothing Comfort* (eds N.R.S. Hollies and R.F. Goldman), Ann Arbor Science Publishers Inc., Michigan, USA, 1977, 121–130.

[37]   Y. Li, J.H. Keighley, and I.F.G. Hampton. Physiological Responses and Psychological Sensations in Wearer Trials with Knitted Sportswear. *Ergonomics*, 1988, **31**, 1709–1721.
[38]   Y. Li, J.H. Keighley, J.E. McIntyre, and I.F.G. Hampton. Predictability Between Objective Physical Factors of Fabrics and Subjective Preference Votes for Derived Garments. *J. Text. Inst.*, 1991, **82**, 277–284.
[39]   A.M. Schneider and B.V. Holcombe. Coolness of 'Cool Wool' Fabrics, *Proc. 8th Int. Wool Textile Res. Conf.*, Christchurch, New Zealand, 1990.
[40]   A.M. Schneider, B.V. Holcombe, and L.G. Stephens. Enhancement of Coolness to the Touch by Hygroscopic Fibers. Part I: Subjective Trials. *Text. Res. J.*, 1996, **66**, 515–520.
[41]   Y. Li. Dimensions of Sensory Perceptions in a Cold Condition. *J. China Text. University,* 1998, Vol. 15, No. 3, 50–53.
[42]   H.M. Elder, S. Fisher, K. Armstrong, and G. Hutchison. Fabric Softness, Handle, and Compression. *J. Text. Inst.*, 1984, **75**, 37–46.
[43]   C. Mackay. *Effect of Laundering on the Sensory and Mechanical Properties of 1 × 1 Rib Knitwear Fabrics*, M.Phil. Thesis, Bolton Institute of Higher Education. Bolton, UK, 1992.
[44]   M.M. Sweeney and D.H. Branson. Sensorial Comfort. Part II: A Magnitude Estimation Approach for Assessing Moisture Sensation. *Text. Res. J.*, 1990, **60**, 447–452.
[45]   C.E. Osgood, G.J. Suci, and P.H. Tannenbaum. *The Measurement of Meaning*, University of Illinois, Urbana, IL, USA, 1957.
[46]   G.A. Kelly. *The Psychology of Personal Constructs*. Norton, New York, NY, USA, 1955.
[47]   H.H. Friedman, L.L. Friedman, and B. Gluck. The Effects of Scale-checking Styles on Responses to a Semantic Differential Scale. *J. Marketing Res. Soc.*, 1988, **30**, 477–481.
[48]   R.H. Evans. The Upgraded Semantic Differential: A Further Test. *J. Market Res. Soc.*, 1980, **2**, 477–481.
[49]   L.L. Golden, G. Albaum, and M. Zimmer. The Numerical Comparative Scale. *J. Retailing*, 1987, **63**, 393–410.
[50]   G. Winakor, C.J. Kim, and L. Wolins. Fabric Hand: Tactile Sensory Assessment. *Text. Res. J.*, 1980, **50**, 601–610.
[51]   P.L. Chen, R.L. Barker, G.W. Smith, and B. Scruggs. Handle of Weft Knit Fabrics. *Text. Res. J.*, 1992, **62**, 200–211.
[52]   A.M. Fritz. A New Way to Measure Fabric Handle. *Text. Asia*, 1992, **23**.
[53]   M.S. Byrne, A.P.W. Garden, and A.M. Fritz. Fiber Types and End-uses: A Perceptual Study. *J. Text. Inst.*, 1993, **84**, 275–288.
[54]   D.P. Bishop. Fabrics: Sensory and Mechanical Properties. *Text. Progress*, 1996, **26**, 1–62.
[55]   N.R. Barnard and A.S.C. Ehrenberg. Robust Measures of Consumer Brand Beliefs. *J. Marketing Res.*, 1990, 477–484.
[56]   J.C. Stevens. Perceived Roughness as a Function of Body Locus. *Perception and Psychophysics*, 1990, **47**, 298–304.
[57]   Y. Li and B.V. Holcombe. Fiber Hygroscopicity and Thermoregulatory Responses During Exercise, presented at *2nd Asian Textile Conf.*, Seoul, South Korea, 1993.
[58]   B.E. Stein and M.A. Meredith. *The Merging of the Senses*. The MIT Press, London, UK, 1993.
[59]   E. Risvik. Understanding Latent Phenomena, in *Multivariate Analysis of Data in Sensory Science* (eds T. Naes and E. Risvik), Elsevier, Amsterdam, The Netherlands, 1996.
[60]   H.M. Elder. Fabric Stiffness. *J. Text. Inst.*, 1984, **75**, 307–311.
[61]   G.A. Miller. The Magical Number Seven, Plus or Minus Two: Some Limits on our Capacity for Processing Information. *Psychological Rev.*, 1956, **63**, 81–97.
[62]   H. Martens. *Determining Sensory Quality of Vegetables, A Multivariate Study*. Agriculture University of Norway, 1986.
[63]   M. Yoshida. The Dimensions of Tactual Impressions (1). *Japanese Psychological Res.*, 1968, **10**, 123–157.
[64]   Y. Li. Predictability Between Human Subjective Preferences and Sensory Factors Towards Clothing During Exercise in a Hot Environment, *26th Textile Research Symp.*, Mt. Fuji, Japan, August 1997.
[65]   Y. Li. Dimensions of Comfort Sensations During Wear in a Hot Condition. Submitted to *J. Federation of Asian Textile Associations*.
[66]   Y. Li. Predictability Between Human Subjective Preferences and Sensory Factors Towards Clothing During Exercise in a Cold Environment. *J. China Text. Uni.*, 1997, Vol. 14, No. 3, 55–60.
[67]   W.W. Cooley and P.R. Lohnes. *Multivariate Data Analysis*. John Wiley & Sons Inc., London, UK, 1971.
[68]   M.M. Tatsuaka. *Multivariate Analysis: Techniques for Educational and Psychological Research*. John Wiley & Sons Inc., London, UK, 1971.
[69]   E. Van der Burg and G. Dijksterhuis. Generalised Canonical Analysis of Individual Sensory Profiles and Instrumental Data, in *Multivariate Analysis of Data in Sensory Science* (eds T. Naes and E. Risvik), Elsevier, New York, NY, USA, 1996.
[70]   Y. Li. Wool for Comfort. *Textile Asia*, Vol. XXV, No. 6, 1994, 57–60.
[71]   M.A. Heller and W. Schiff. *The Psychology of Touch*, Lawrence Erlbaum Associates, Hove and London, UK, 1991, 354.
[72]   S. Coren and L.M. Ward. *Sensation and Perception*. Harcourt Brace Jovanovich, New York, NY, USA, 1989.
[73]   A. Iggo. Sensory Receptors, Cutaneous, in *Sensory Systems II: Senses other than Vision* (ed J.M. Wolfe), A Pro Scientia Viva Title, Boston, USA, 1988, 109–110.

[74]   T. Amano, C. Minakuchi, and K. Takada. Spectrum Analysis of Clothing Pressure Fluctuation. *Sen-i Gakkaishi*, 1996, **52**, 41–44.

[75]   S. Sukigara and T. Ishibashi. Analysis of Frictional Properties Related to Surface Roughness of Crepe Fabric. *Sen-i Gakkaishi*, 1994, **50**, 349–356.

[76]   H. Momota, H. Makabe, T. Mitsuno, and K. Ueda. A Study of Clothing Pressure Caused by Japanese Women's High Socks. *J. Japan Res. Association for Text. End-uses*, 1993, **34**, 603–614.

[77]   H. Momota, H. Makabe, T. Mitsuno, and K. Ueda,. A Study of Clothing Pressure Caused by Japanese Men's Socks. *J. Japan Res. Association for Text. End-uses*, 1993, **34**, 175–186.

[78]   H. Makabe, H. Momota, T. Mitsuno, and K. Ueda. Effect of Covered Area at the Waist on Clothing Pressure. *Sen-i Gakkaishi*, 1993, **49**, 513–521.

[79]   H. Shimizu. Dynamic Measurement of Clothing Pressure on the Body in a Brassiere. *Sen-i Gakkaishi*, 1993, **49**(1), 57–62, **49**, 57–62, 1993.

[80]   H. Makabe, H. Momota, T. Mitsuno, and K. Ueda. A Study of Clothing Pressure Developed by the Brassiere. *J. Japan Res. Assoc. for Text. End-uses*, 1991, **32**, 416–423.

[81]   H. Shimizu, U. Totsuka, and Y. Shimizu. Dynamic Behavior of Clothing Pressure on the Body in Slacks. *Sen-i Gakkaishi*, 1990, **46**, 237–243.

[82]   R.A. Westerman, R.K. Garnsworthy, A. Walker, P. Kenins, R.L. Gully, and P. Fergin. Aspects of Human Cutaneous Small Nerve Function: Sensations of Prickle and Itch, presented at *29th IUPS Satellite Symp.*, Budapest, Hungary, 1984.

[83]   R.K. Garnsworthy, R.L. Gully, P. Kenins, and R.A. Westerman. Transcutaneous Electrical Stimulation and the Sensation of Prickle. *J. Neurophysiology.*, 1988, **59**, 1116–1127.

[84]   R.K. Garnsworthy, R.L. Gully, P. Kenins, R.J. Mayfield, and R.A. Westerman. Identification of the Physical Stimulus and the Neural Basis of Fabric-evoked Prickle. *J. Neurophysiology*, 1988, **59**, 1083–1097.

[85]   P. Kenins. The Functional Anatomy of the Receptive Fields of Rabbit C Polymodal Nociceptors. *J. Neurophysiology*, 1988, **59**, 1098–1115.

[86]   R.K. Garnsworthy, R.L. Gully, R.P. Kandiah, P. Kenins, R.J. Mayfield, and R.A. Westerman. Understanding the Causes of Prickle and Itch from the Skin Contact of Fabrics. *Australian Text.*, 1988, **8**, 26–29.

[87]   W.D. Willis. *The Pain System: The Neural Basis of Nociceptive Transmission in the Mammalian Nervous System.* Karger, Basel, Switzerland, **8**, 1985.

[88]   F. Lembeck and R. Gamse. Substance P in Peripheral Sensory Processes, presented at *Ciba Foundation Symp. 91*, London, UK, 1982.

[89]   F. Lembeck. Sir Thomas Lewis's Nocicensor System, Listamine and Substance-P-containing Primary Afferent Nerves. *TINS*, 1983, 106–108.

[90]   S.J. Lederman and M.M. Taylor. Fingertip Force, Surface Geometry, and the Perception of Roughness by Active Touch. *Perception and Psychophysics*, 1972, **12**, 401–408.

[91]   S.J. Lederman. Tactile Roughness of Grooved Surfaces: the Touching Process and Effects of Macro- and Micro-structure. *Perception and Psychophysics*, 1974, **16**, 385–395.

[92]   S.J. Lederman. The Perception of Surface Roughness by Active and Passive Touch. *Bull. of the Psychonomic Soc.*, 1981, **18**, 498–511.

[93]   S.J. Lederman. Tactual Roughness Perception: Spatial and Temporal Determinants. *Canadian J. Psychology*, 1983, **37**, 498–511.

[94]   I. Darian-Smith and L. Oke. Peripheral Neural Representation of the Spatial Frequency of a Grating Moving Across the Monkey's Finger Pad. *J. Physiology*, 1980, **309**, 117–133.

[95]   L. Kruger. *Pain and Touch*, 2nd ed. Academic Press, New York, NY, USA, 1996.

[96]   A.W. Goodwin and J.W. Morley. Sinusoidal Movement of Grating Across the Monkey's Fingerpad: Effect of Contact Angle and Force of the Grating on Afferent Fiber Response. *J. Neuroscience*, 1987, **7**, 2192–2202.

[97]   A.W. Goodwin and J.W. Morley. Sinusoidal Movement of a Grating Across the Monkey's Fingerpad: Representation of Grating and Movement Features in Afferent Fiber Response. *J. Neuroscience*, 1987, **7**, 2168–2180.

[98]   A.W. Goodwin, K.T. John, K. Sarhian, and I. Darian-Smith. Spatial and Temporal Factors Determining Afferent Fiber Responses to a Grating Moving Sinusoidally over the Monkey's Fingerpad. *J. Neuroscience*, 1989, **9**, 1280–1293.

[99]   J.D. Greenspan and S.J. Bolanowski. The Psychophysics of Tactile Perception and its Peripheral Physiological Basis. In *Pain and Touch* (ed L. Kruger), Academic Press, 1996, London, UK, 25–104.

[100]   C.E. Connor, S.S. Hsiao, J.R. Phillips, and K.O. Johnson. Tactile Roughness: Neural Codes that Account for Psychophysical Magnitude Estimates. *J. Neuroscience*, 1990, **10**, 3823–3836.

[101]   C.E. Connor and K.O. Johnson. Neural Coding of Tactile Texture: Comparison of Spatial and Temporal Mechanisms for Roughness Perception. *J. Neuroscience*, 1992, **12**, 3414–3426.

[102]   K.O. Johnson and S.S. Hsiao. Evaluation of the Relative Roles of Slowly and Rapidly Adapting Afferent Fibers in Roughness Perception. *Canadian J. Physiology Pharmacology*, 1994, **72**, 488–497.

[103]   R.H. LaMotte. Psychophysical and Neurophysiological Studies of Tactile Sensitivity, in *Clothing Comfort* (eds N.R.S. Hollies and R.F. Goldman), Ann Arbor Science Publishers, Michigan, USA, 1977, 85–105.

[104]   A.R. Gwosdow, J.C. Stevens, L.G. Berglund, and J.A.J. Stolwijk. Skin Friction and Fabric Sensations in Neutral and Warm Environments. *Text. Res. J.*, 1986, **56**, 574–580.

[105] F.W. Behmann. Tests on the Roughness of Textile Surfaces. *Melliand Textilber.*, 1990, **71**, 438–440 and E199–200.

[106] C.A. Wilson and R.M. Laing. The Effect of Wool Fiber Variables on Tactile Characteristics of Homogeneous Woven Fabrics. *Cloth. Text. Res. J.*, 1995, **13**, 208–212.

[107] D.G. Mehrtens and K.C. McAlister. Fiber Properties Responsible for Garment Comfort. *Text. Res. J.*, 1962, **32**, 658–665.

[108] K.L. Harter, S.M. Spivak, K. Yeh, and T.L. Vigo. Applications of the Trace Gas Technique in Clothing Comfort. *Text. Res. J.*, 1981, 345–355.

[109] D.C. Spray. Thermoreceptors, in *Sensory Systems II: Senses other than Vision* (ed. J.M. Wolfe), A Pro Scientia Viva Title, Basel, Switzerland, 1988, 132–133.

[110] H. Hensel. *Thermoreception and Temperature Regulation.* Academic Press, New York, NY, USA, 1981.

[111] K. Ring and R.de Dear. Temperature Transients: A Model for Heat Diffusion Through the Skin: Thermoreceptor Responses and Thermal Sensations. *Indoor Air*, 1991, **4**, 448–456.

[112] H. Hensel and D.R. Kenshalo. Warm Receptors in the Nasal Region of Cats. *J. Physiology*, 1969, **204**, 99–112.

[113] D.R. Kenshalo. Correlations of Temperature Sensitivity in Man and Monkey, a First Approximation, in *Sensory Functions of the Skin* (ed. Zotterman), Pergamon Press, Oxford, UK, and New York, NY, USA, 1976, 305–330.

[114] R.W. Dykes. Coding of Steady and Transient Temperatures by Cutaneous 'Cold' Fibers Serving the Hand of Monkeys. *Brain Research*, 1975, **98**, 485–500.

[115] Y. Li, B.V. Holcombe, and R. de Dear. Enhancement of Coolness to the Touch by Hygroscopic Fibers. Part II: Physical Mechanisms. *Text. Res. J.*, 1996, **66**, 587–595.

[116] R. Nielsen and T.L. Endrusick. Sensations of Temperature and Humidity during Alternative Work/Rest and the Influence of Underwear Knit Structure. *Ergonomics*, 1990, **33**, 221–234.

[117] M.M. Sweeney and D.H. Branson. Sensorial Comfort. Part I: A Psychological Method for Assessing Moisture Sensation in Clothing. *Text. Res. J.*, 1990, **60**, 371–337.

[118] R.P. Clark and O.G. Edholm. *Man and his Thermal Environment.* Edward Arnold, 1985.

[119] P. Kenins, J. Spence, and C.J. Veitch. *Human Perception of Air Humidity.* Laboratory Report GL33, CSIRO, Division of Wool Technology, Geelong, Australia, Feb. 1992.

[120] I.M. Bentley. The Synthetic Experiment. *American J. Psychology*, 1900, **11**, 405–425.

[121] Y. Li, A.M. Plante, and B.V. Holcombe. Fiber Hygroscopicity and Perceptions of Dampness. Part II: Physical Mechanisms. *Text. Res. J.*, 1995, **65**, 316–324.

[122] A.M. Fiore and P.A. Kimle. *Understanding Aesthetics for the Merchandising and Design Professional.* Fairchild Publications, New York, NY, USA, 1997.

[123] S.J. Lederman. The Perception of Texture by Touch, in *Tactual Perception: A Source Book* (eds W. Schiff and E. Foulke), Cambridge University Press, Cambridge, UK, 1982, 130–167.

[124] D.M. Brown, B. Cameron, L.D. Burns, M.J. Dallas, J. Chandler, and S. Kaiser. The Effect of Sensory Interaction on Descriptions of Fabrics, *Proc. of the Ann. Meeting Int. Textile & Apparel Assoc. Inc.*, 1994.

[125] M.J. Dallas, B. Brandt, R. Smitley, and C. Salusso. Sensory Perceptions of Fabric: Gender Differences, *Proc. Ann. Meeting of the Int. Textile & Apparel Assoc. Inc.*, 1994.

[126] C. Jung and N. Naruse. Effects of Difference in Weave on the Visual Evaluation of White Silk Fabrics and Black Ones. *Sen-i Gakkaishi*, 1995, **51**(7), 338–344.

[127] S.J. Kadolph, A.L. Langford, N. Hollen, and J. Saddler. *Textiles.* MacMillan, New York, NY, USA, 1993.

[128] A.B. Vallbo. Touch, Sensory Coding in the Human Hand, in *Sensory Systems II: Senses other than Vision* (ed. J.M. Wolfe), A Pro Scientia Viva Title, Boston, USA, 1988, 136–138.

[129] J.J. Gibson. Observations on Active Touch. *Psychological Rev.*, 1962, **69**, 477–490.

[130] D. Katz. *The World of Touch.* Lawrence Erlbaum Associates, Hillsdale, NJ, USA, 1989.

[131] P.W. Davidson, S. Abbot, and J. Gershenfeld. Influence of Exploration Time on Haptic and Visual Matching of Complex Shape. *Perception and Psychophysics*, 1974, **15**, 539–543.

[132] P.W. Davidson. Some Functions of Active Handling: Studies with Blind Humans, *New Outlook for the Blind*, 1976, **70**, 198–202.

[133] S.J. Lederman and R.L. Klatzky. Hand Movements: A Window into Haptic Object Recognition. *Cognitive Psychology*, 1987, **19**, 342–368.

[134] H.M. Elder, S. Fisher, G. Hutchison, and S. Beattie. A Psychological Scale for Fabric Stiffness. *J. Text. Inst.*, 1985, **76**, 442.

[135] J.L. Hu, W.X. Chen, and A. Newton. A Psychophysical Model for Objective Hand Evaluation: An Application of Stevens' Law. *J. Text. Inst.*, 1993, **84**, 354–363.

[136] C.P. Yaglou and W.E. Miller. Effective Temperature with Clothing. *Trans. Amer. Soc. Heat Vent. Egn. Assoc.*, 1925, **31**, 89.

[137] F.H. Rohles. The Measurement and Prediction of Thermal Comfort. *ASHRAE*, 1967, **118**, 98–114.

[138] A.P. Gagge, J.A.J. Stolwijk, and J.D. Hardy. Comfort and Thermal Sensations and Associated Physiological Responses at Various Ambient Temperatures. *Environmental Res.*, 1967, 1–20.

[139] A.P. Gagge, J.A.J. Stolwijk, and B. Saltin. Comfort, and Thermal Sensations and Associated Physiological Responses during Exercise at Various Ambient Temperatures. *Environmental Res.*, 1969, 209–229.

[140] E.C. Carterete and M.P. Friedman. *Handbook of Perception: Psychological Judgement and Measurement*, vol. 2. Academic press, London, UK, 1973.

[141] E.C. Carterete and M.P. Friedman. *Handbook of Perception: Biology of Perception System*. Academic Press, London, UK, 1974.
[142] P.E. McNall, J. Jaax, F.H. Rohles, and W.E. Springer. Thermal Comfort (Thermal Neutral) Conditions for Three Levels of Activity. *ASHRAE Transactions*, 1967, **73**.
[143] P.O. Fanger. Thermal Environment – Human Requirements. *Sulzer Technical Rev.*, 1985, **67**, 3–6.
[144] A.P. Gagge. Rational Temperature Indices of Man's Thermal Environment and their use with a 2-node Model of his Temperature Regulation. *Fed. Proc.*, 1973, **32**, 1572–1582.
[145] H. Hensel. Neural Processes in Thermoregulation. *Physiol. Rev.*, 1973, **53**, 948–1017.
[146] A.P. Gagge and Y. Nishi. Heat Exchange between Human Skin Surface and Thermal Environment, in *Handbook of Physiology* (ed. D.H.K. Lee), American Physiology Society, Bethesda, Md, USA, 1977, 69–92.
[147] D. Mitchell, A.R. Atkins, and C.H. Wyndham. Mathematical and Physical Models of Thermoregulation, in *Essays on Temperature Regulation* (eds J. Bligh and R.E. Moore), North-Holland Publication, Amsterdam and London, 1972, 37–54.
[148] J. Wenner. Mathematical Treatment of Structure and Function of the Human Thermoregulatory System. *Biological Cybernetics*, 1977, **25**, 93–101.
[149] J.S. Hayward, J.D. Eckerson, and M.L. Collis. Thermoregulatory Heat Production in Man: Prediction Equation based Skin and Core Temperatures. *J. Appl. Physiol.*, 1977, **42**, 377–384.
[150] J.A.J. Stolwijk and J.D. Hardy. Control of Body Temperature, in *Handbook of Physiology*, Vol. 9 (eds. D.H.K. Lee), American Physiology Society, Bethesda, Md, USA, 1977.
[151] H.G. David. The Buffering Action of Hygroscopic Clothing. *Text. Res. J.*, 1964, **34**, 814–816.
[152] I.M. Stuart, A.M. Schneider, and T.R. Turner. Perception of the Heat of Sorption of Wool. *Text. Res. J.*, 1989, **59**, 324–329.
[153] R.J. de Dear, H.N. Knudsen, and P.O. Fanger. Impact of Air Humidity on Thermal Comfort during Step-changes. *ASHRAE Trans.*, 1989, **95**, 336–350.
[154] A. Shitzer and J.C. Chato. Thermal Interaction with Garments, in *Heat Transfer in Medicine and Biology, Volume 1: Analysis and Applications*, (eds. A. Shitzer and R.C. Eberhart) Plenum Press, New York, NY, USA, 1985.
[155] B.W. Jones, M. Ito, and E.A. McCullough. Transient Thermal Responses of Clothing Systems, *Proc. Int. Conf. Environ. Ergonomics-IV*, Austin, Texas, USA, 1990.
[156] B.W. Jones and Y. Ogawa. Transient Response of the Human–clothing System, *Proc. Int. Conf. on Human–Environment System*, Tokyo, Japan, 1991.
[157] Y. Li and B.V. Holcombe. Mathematical Simulation of Heat and Mass Transfer in Human–Clothing–Environment System. *Text. Res. J.*, 1998, **67**, 5, 389–397.
[158] Y. Li. The Buffering Effect of Hygroscopic Clothing Against Rain, presented at *The 4th Asian Textile Conf.*, Taipei, Taiwan, 1997.
[159] K. Slater. Comfort Properties of Textiles. *Text. Progress*, 1977, **9**, 1–91.
[160] A.H. Woodcock. Moisture Transfer in Textile Systems. Part I. *Text. Res. J.*, 1962, **32**, 628–633.
[161] A.P. Gagge, A.C. Burton, and H.C. Bazett. A Practical System of Units for the Description of the Heat Exchange of Man with his Environment. *Science*, 1941, **94**, 428–430.
[162] J.H. Mecheels and K.H. Umbach. The Psychrometric Range of Clothing Systems, in *Clothing Comfort: Interaction of Thermal, Ventilation, Construction and Assessment Factors* (eds N.R.S. Hollies and R.F. Goldman), Ann Arbor Science Publishers Inc., Michigan, USA, 1977.
[163] J.R. Breckenridge. Effects of Body Motion on Convective and Evaporative Heat Exchanges through Various Design of Clothing, in *Clothing Comfort: Interaction of Thermal, Ventilation, Construction and Assessment Factors* (eds N.R.S. Hollies and R.F. Goldman), Ann Arbor Science Publishers Inc., Michigan, USA, 1977.
[164] J. Mecheels. Die Messung der Functionellen Wirkung der Kleidung auf den Menschen. *Melliand Textilber.*, 1971, **52**.
[165] P.S.H. Henry. Diffusion in Absorbing Media. *Proc. Roy. Soc.*, 1939, **171** A, 215–241.
[166] P.S.H. Henry. The Diffusion of Moisture and Heat through Textiles. *Discussions of the Faraday Soc.*, 1948, **3**, 243–257.
[167] D.W. Lyons and C.T. Vollers. *Text. Res. J.*, 1971, **41**, 661–668.
[168] Y. Li. *Liquid Transport and Active Sportswear*, CSIRO, Division of Wool Technology, Geelong, Australia, WTC97.01, 1996.
[169] P. Gibson. *Governing Equations for Multiphase Heat and Mass Transfer in Hygroscopic Porous Media with Applications to Clothing Materials*, Army Natick Research, Development and Engineering Center, Natick, Massachusetts 01760-5000, USA, Technical Report NATICK/TR-95/004, November, 1994.
[170] J. Crank. *The Mathematics of Diffusion*. Clarendon Press, Oxford, UK, 1975.
[171] H.G. David and P. Nordon. Case Studies of Coupled Heat and Moisture Diffusion in Wool Beds. *Text. Res. J.*, 1969, **39**, 166–172.
[172] B. Farnworth. A Numerical Model of the Combined Diffusion of Heat and Water Vapor through Clothing. *Text. Res. J.*, 1986, **56**, 653–665.
[173] J.A. Wehner. *Moisture Transport through Fiber Networks*, in Textile Research Institute and Department of Chemical Engineering, Ph.D Thesis, Princeton University, USA, 1987.

[174] J.G. Downes and B.H. Mackay. Sorption Kinetics of Water Vapor in Wool Fibers. *J. Poly. Sci.*, 1958, **28**, 45–67.

[175] I.C. Watt. Kinetic Studies of the Wool–Water System. Part I: The Influence of Water Concentration. *Text. Res. J.*, 1960, **30**, 443–450.

[176] I.C. Watt. Kinetic Studies of the Wool–Water System. Part II: The Mechanisms of Two-stage Absorption. *Text. Res. J.*, 1960, **30**, 644–651.

[177] Y. Li and B.V. Holcombe. A Two-stage Sorption Model of the Coupled Diffusion of Moisture and Heat in Wool Fabrics. *Text. Res. J.*, 1992, **62**, 211–217.

[178] P. Nordon and H.G. David. Coupled Diffusion of Moisture and Heat in Hygroscopic Textile Materials. *Int. J. Heat Mass Transfer*, 1967, **10**, 853–866.

[179] Y. Li, B.V. Holcombe, A.M. Schneider, and F. Apcar. Mathematical Modelling of the Coolness to Touch of Hygroscopic Fabrics. *J. Text. Inst.*, 1993, **84**, 267–274.

[180] A. Rae and B. Rollo. *The WIRA Textile Data Book.* WIRA, Leeds, UK, 1973.

[181] A.R. Urguhart and A.M. Williams. The Moisture Relations of Cotton: the Effect of Temperature on the Absorption of Water by Soda-boiled Cotton. *J. Text. Inst.*, 1924, **15**, T559–T572.

[182] A.M. Schneider. Heat Transfer through Moist Fabrics. University of New South Wales, 1987, Sydney, Australia, Ph.D Thesis.

[183] P. Wagman *et al. American Institute of Physics Handbook*, 2nd ed: American Institute of Physics, USA, 1975.

[184] Y. Li and B.V. Holcombe. *A Kinetic Approach Modelling the Coupled Diffusion of Moisture and Heat in Wool Fabrics*, CSIRO Division of Wool Technology, Sydney, Australia, Laboratory Note SN/125, 1991.

[185] Y. Li and Z.X. Luo. An Improved Two-stage Sorption of the Coupled Diffusion of Moisture and Heat in Wool Fabrics, *Text. Res. J.*, 1999, 760–768.

[186] Y. Li and Z.X. Luo. Physical Mechanisms of Moisture Transfer in Hygroscopic Fabrics under Humidity Transients, Accepted by *J. Text. Inst.*, 1998.

[187] A.M. Schneider and B.V. Holcombe. Properties Influencing Coolness to the Touch of Fabrics. *Text. Res. J.*, 1991, **61**, 488–494.

[188] Y. Li and G. Brown. *Fabric Coolness-to-the-touch. Part III: Relationships between Measured Coolness and Fabric Fundamental Properties.* CSIRO, Division of Wool Technology, Ryde, Sydney, Australia, Report WTC96.23, May, 1996.

[189] S. Kawabata. Development of a Device for Measuring Heat–Moisture Transfer Properties of Apparel Fabrics. *J. Text. Mach. Soc. Japan*, 1984, **37**, 38–49.

[190] M. Yoneda and S. Kawabata. Analysis of Transient Heat Conduction and its Applications. Part I: The Fundamental Analysis and Applications to Thermal Conductivity and Thermal Diffusivity Measurements. *J. Text. Mach. Soc. Japan*, 1983, **29**, 73–83.

[191] M. Yoneda and S. Kawabata. Analysis of Transient Heat Conduction and its Applications. Part II: Theoretical Analysis of the Relationship between Warm/Cool Feeling and Transient Heat Conduction in Skin. *J. Text. Mach. Soc. Japan*, 1985, **31**, 79–85.

[192] L. Hes, I. Dolezal, and J. Handzl. New Method and Facility for the Objective Evaluation of Thermal Contact Properties of Textile Fabrics. *Melliand Textilber.*, 1990, **76**, 679.

[193] Y. Li. *Fabric Coolness-to-the-touch. Part II: Methods of Measurements*, CSIRO, Division of Wool Technology, Ryde, Sydney, Australia, Report WTC 96.22, May, 1996.

[194] H.G. David. The Effect of Changing Humidity on the Insulation of Hygroscopic Clothing. *Text. Res. J.*, 1965, **35**, 820–826.

[195] B.W. Olesen and R. Nielsen. *Thermal Insulation of Clothing Measured on a Movable Thermal Manikin and Human Subjects*, Technical University of Denmark, Technical Report 7206/00/914, 1983.

[196] R.J. de Dear, S. Tanabe, H.N. Knudsen, J. Pejtersen, J. Mackeprang, and P.O. Fanger. Thermal Responses of Wool Clothing during Humidity Transients, presented at *Indoor Air Conf.*, Berlin, Germany, 1987.

[197] J.F. Mackeprang, H.N. Knudsen, and P.O. Fanger. The Physiological Impact of Sorption Heat in Hygroscopic Clothing. Presented at *Int. Conf. on Environmental Ergonomics – IV*, Texas, USA, 1990.

[198] D.M. Scheurell, S.M. Spivak, and N.R.S. Hollies. Dynamic Surface Wetness of Fabrics in Relation to Clothing Comfort. *Text. Res. J.*, 1985, 394–399.

[199] K. Hong, N.R.S. Hollies, and S.M. Spivak. Dynamic Moisture Vapor Transfer Through Textiles. Part I: Clothing Hygrometry and the Influence of Fiber Properties. *Text. Res. J.*, 1988, **57**, 697–706.

[200] C.W. Hock, A.M. Sookne, and M. Harris. Thermal Properties of Moist Fabrics. *J. Res. Natl. Bur. Stand.*, 1944, **32**, 229–252.

[201] A.B.D. Cassie, B.E. Atkins, and G. King. Thermo-static Action of Textile Fibers. *Nature*, 1939, **143**, 162.

[202] A.B.D. Cassie. Fibers and Fluids. *J. Text. Inst.*, 1962, **53**, P739–P745.

[203] E.C. Rodwell, E.T. Renbourn, J. Greenland, and K. Kenchington. An Investigation of the Physiological Value of Sorption Heat in Clothing Assemblies. *J. Text. Inst.*, 1965, **56**, T624–T645.

[204] K.H. Umbach. Hautnahe Synthetics mit Gutem Tragekomfort. *Chemiefasern/Textilind.*, 1980, **30/82**, 628–636.

[205] J.H. Andreen, J.W. Gibson, and O.C. Wetmore. Fabric Evaluations Based on Physiological Measurements of Comfort. *Text. Res. J.*, 1953, **23**, 11–22.

[206] Z. Vokac, V. Kopke, and P. Keul. Physiological Responses and Thermal, Humidity, and Comfort Sensations in Wear Trials with Cotton and Polypropylene Vests. *Text. Res. J.*, 1976, **46**, 30–38.

[207] I. Holmer. Heat Exchange and Thermal Insulation Compared in Woolen and Nylon Garments during Wear Trials. *Text. Res. J.*, 1985, **55**, 511–518.

[208] K.L. Hatch, N.L. Markee, H.L. Maibach, R.L. Barker, S.S. Woo, and P. Radhakrishnaiah. Viva Cutaneous and Perceived Comfort Responses to Fabric. Part III: Water and Blood Flow in Human Skin under Garments Worn by Exercising Subjects in a Hot, Humid Environment. *Text. Res. J.*, 1990, **60**, 510–519.

[209] J.L. Spencer-Smith. Physical Basis of Clothing Comfort. Part V: The Behaviour of Clothing in Transient Conditions. *Clothing Res. J.*, 1978, 21–30.

[210] J.A. Wehner, B. Miller, and L. Rebenfeld. Dynamics of Water Vapor Transmission through Fabric Barriers. *Text. Res. J.*, 1988, **58**.

[211] F.W. Behmann. Influence of the Sorption Properties of Clothing on Sweating Loss and the Subjective Feeling of Sweating. *Appl. Polym. Symp.*, 1971, **18**, 1477–1482.

[212] H. Tokura, Y. Yamashita, and S. Tmioka. The Effect of Moisture and Water Absorbency of Fibers on the Sweating Rates of Sedentary Man in Hot Ambient, in *Objective Specification of Fabric Quality, Mechanical Properties and Performance* (eds S. Kawabata, R. Postle, and M. Niwa), Text. Mach. Soc. of Japan, 1982, 407–418.

[213] M. Matsudaira, J.D. Watt, and G.A. Carnaby. Measurement of the Surface Prickle of Fabrics. Part I: The Evaluation of Potential Objective Methods. *J. Text. Inst.*, 1990, **81**, 288–299.

[214] C.J. Veitch and G.R.S. Naylor. The Mechanics of Fiber Buckling in Relation to Fabric-evoked Prickle. *Wool Technol. Sheep Breeding*, 1992, **40**, 31–34.

[215] G.R.S. Naylor. The Role of Coarse Fibers in Fabric Prickle using Blended Acrylic Fibers of Different Diameters. *Wool Technol. Sheep Breeding*, 1992, **40**, 14–18.

[216] M. Matsudaira, J.D. Watt, and G.A. Carnaby. Measurement of the Surface Prickle of Fabrics. Part II: Some Effects of Finishing on Fabric Prickle. *J. Text. Inst.*, 1990, **81**, 300–309.

[217] G.R.S. Naylor, D.G. Phillips, and C.J. Veitch. Fabric-evoked Prickle in Worsted Spun Single Jersey Fabrics. Part I: The Role of Fiber end Diameter Characteristics. *Text. Res. J.*, 1997, **67**, 288–295.

[218] P. Kenins. The Cause of Prickle and the Effect of some Fabric Construction Parameters on Prickle Sensations. *Wool Technol. Sheep Breeding*, 1992, **40**, 19–24.

[219] Y. Li and J. Keighley. Relations between Fiber, Yarn, Fabric Mechanical Properties, and Subjective Sensory Responses in Wear Trials. Presented at *The 3rd Int. Conf. on Ergonomics*, Helsinki, Finland, 1988.

[220] Y. Li. *The Objective Assessment of Comfort of Knitted Sportswear in Relation to Psycho–Physiological Sensory Studies*, Dept. of Textile Industries, The Univ. of Leeds, Leeds, UK, 1988, Ph.D Thesis.

[221] H.M. Elder, S. Fisher, K. Armstrong, and G. Hutchison. Fabric Stiffness, Handle, and Flexion. *J. Text. Inst.*, 1984, **75**, 99–106.

[222] F.T. Peirce. The Handle of Cloth as a Measureable Quantity. *J. Text. Inst.*, 1930, **21**, T377.

[223] N. Ito. The Relation Among the Biaxial Extension Properties of Girdle. *Sen-i Seihin Shohi Kagaku*, 1995, **36**, 102–109.

[224] J.O. Ajayi. Fabric Smoothness Friction and Handle. *Text. Res. J.*, 1992, **62**, 52–59.

[225] R.B. Ramgulam, J. Amirbayat, and I. Porat. Measurement of Fabric Roughness by a Non-contact Method. *J. Text. Inst.*, 1993, **84**, 99–106.

[226] W.J. Kirk and S.M. Ibrahim. Fundamental Relationship of Fabric Extensibility to Anthropometric Requirements and Garment Performance. *Text. Res. J.*, 1966, **57**, 37–47.

[227] M.J. Denton. Fit, Stretch, and Comfort. Presented at *3rd Shirley Int. Seminar: Textiles for Comfort*, Manchester, UK, 1970.

[228] J. Lemmens. *Industr. Text. Belge.*, 1966, **8**, 71.

[229] S.M. Ibrahim. Mechanics of Form-persuasive Garments Based on Spandex Fibers. *Text. Res. J.*, 1968, **38**, 950.

[230] E.M. Growther. Comfort and Fit in 100% Cotton-Denim Jeans. *J. Text. Inst.*, 1985, 323–338.

[231] M. Rutten. *Zeissdinft Orthop. & Ihre Ereuzgebiete*, 1978, **16**, 176.

[232] T.K. Davidson. Panty-girdle Syndrome. *Br. Med. J.*, 1972, 407.

[233] S. Kawabata. The Development of the Objective Measurement of Fabric Handle, *Proc. of 1st Japan–Australia Symp. on Objective Specification of Fabric Quality, Mechanical Properties, and Performance*, Osaka, Japan, 1982.

[234] S. Kawabata. Characterisation Method of the Physical Properties of Fabrics and the Measuring System for Hand-feeling Evaluation. *J. Text. Mach. Soc., Japan*, 1973, **26**, 721–728.

[235] K. Slater. Physical Testing and Quality Control. *Text. Progress*, 1993, 23, **1**, 1–171.

[236] T.J. Mahar, R.C. Dhingra, and R. Postle. Measuring and Interpreting Low-stress Fabric Mechanical and Surface Properties. *Text. Res. J.*, 1987, **57**, 357.

[237] S. Kawabata and M. Niwa. Objective Measurement of Fabric Mechanical Properties and Quality. *Int. J. Cloth. Sci. Technol.*, 1991, **3**, 7.

[238] K.H. Umbach. Methods of Measurement for Testing Physiological Requirements of Civilian, Work and Protective Clothing and Uniforms. *Melliand Textilber.*, 1987, **68**, 857–865.

[239] L. Vollrath and H. Martin. Relationships between the Sensory Judgement of the Skin Contact Behaviour of Fabrics and Laboratory Testing of Properties. *Textiltechnik*, 1983, 225–231.

[240] I. Creig. The Application of a New Generation of Consumer Research Techniques to Predicting and Maximising Market Share in the Fiber Market. *J. Text. Inst.*, 1994, **85**, 231–243.

[241] R. Dayal. Comfort Properties of Textiles. Part I. *Text. Dyer Printer*, 1980, **49**, 21–24.

[242] K. Greenwood. Making Clothing more Comfortable. *Text. Month*, 1971, 107.

[243] ASTM. *Manual on Sensory Testing Methods*, American Society for Testing and Materials ASTM Special Technical Publication 434, 1986.

[244] J.E. Lyn. Comfort, it's More than Meets the Eye. *Amer. Dyest. Rep.*, 1978, **67**, 9.

[245] R.N. DeMartino. Comfort Properties of Polybenzimidazole Fiber. *Text. Res. J.*, 1984, 516–521.

[246] N.R.S. Hollies. Visual and Tactile Perceptions of Textile Quality. *J. Text. Inst.*, 1989, **80**, 1–18.

[247] P. Kotler. *Marketing Management*, 8th ed. Prentice Hall International Inc., London, UK, 1994.

[248] R.G. Cooper and E.J. Kleinschmidt. New Product Processes at Leading Industrial Firms. *Industrial Marketing Management*, 1991, **20**, 137–147.

[249] R.G. Cooper and E.J. Kleinschmidt. *New Products: The Key Factors In Success*. American Marketing Association, Chicago, USA, 1990.

[250] K.U. Umbach. Protective Clothing Against Cold with a Wide Range of Thermophysiological Control. *Melliand Textilber.*, 1981, **3** and **4**, 360–364, 456–462 (English ed.).

[251] R. Vatsala and V. Subramaniam. The Integral Evaluation of Fabric Performance. *J. Text. Inst.*, 1993, **84**, 495–500.

[252] Y. Li. Wool Sensory Properties and Product Development, *Textile Asia*, 1998, Vol. XXIX, **5**, 35–39.

NAME INDEX*

* This name index is based on reference numbers, not page numbers.

TITLES CURRENTLY AVAILABLE FROM THE TEXTILE PROGRESS SERIES

For further information on how to order back copies of Textile Progress or about subscriptions, please contact:

Subscriptions Dept, The Textile Institute International, Fourth Floor, St. James's Buildings, Oxford Street, Manchester M1 6FQ, UK.

Tel: +44(0)161 237 1188, Fax: +44(0)161 236 1991, Email: tiihq@textileinst.org.uk, website: http://www.texi.org